MW00856856

The Essential Guide to Our Night Sky

SOUTHERN CALIFORNIA

STARWATCH

Make the stars
your old friends!

Mike
Lynch

VOYAGEUR PRESS

DEDICATED TO MY WIFE, KATHY; MY MOTHER, EILEEN; MY CHILDREN, ANGIE AND SHAUN; AND DEAR FRIEND TANJA ALLARD, WHO HAVE ALL SUPPORTED ME IN WRITING THIS BOOK.

ALSO A SPECIAL DEDICATION TO MY DAD, COLONEL DON LYNCH, WHO LEFT THIS WORLD AND HAS TAKEN THE ULTIMATE TRIP TO THE HEAVENS.

Be glad of life because it gives you the chance to love and to work and to play and to look up at the stars.

—Henry Van Dyke

FRONT COVER, TOP: Boy viewing comet Hale Bopp (Photograph © Richard Hamilton Smith/ DPA Photo)

FRONT COVER, BOTTOM ROW, LEFT TO RIGHT: Whirlpool Galaxy (Photograph © Robert Gendler)

The Horsehead Nebula (Photograph © Robert Gendler)

Pleiades (Photograph © Robert Gendler)

BACK COVER: Mike Lynch (Photograph © Kathy Lynch)

FRONTISPIECE: Milky Way (Photograph © Thomas Matheson)

TITLE PAGE: Pleiades, M45, in the constellation Taurus (Photograph © Robert Gendler)

FACING PAGE: North America Nebula, NGC7000, in the constellation Cygnus (Photograph © Thomas Matheson)

First published in 2007 by Voyageur Press, an imprint of MBI Publishing Company, Galtier Plaza, Suite 200, 380 Jackson Street, St. Paul, MN 55101-3885 USA

Library of Congress Cataloging-in-Publication Data

Lynch, Mike, 1956-
Southern California starwatch : the essential guide to our night sky / Mike Lynch.
p. cm.
Includes bibliographical references and index.
ISBN-13: 978-0-7603-2841-5 (concealed wire o bound)
ISBN-10: 0-7603-2841-2 (concealed wire o bound)
1. Astronomy--California, Southern--Observers' manuals. I. Title.
QB64.L9655 2007
522.09794'9--dc22
2006025408

Edited by Danielle J. Ibister
Designed by Julie Vermeer
Constellation illustrations by John Tapp

Printed in China

CONTENTS

FACING PAGE: Whirlpool Galaxy, M51, in the constellation Ursa Major (Photograph © Robert Gendler)

INTRODUCTION

Make the Stars Your Old Friends

Why do you need a stargazing book specific to Southern California? Because the night sky looks different depending on where you are stargazing. Did you know, for example, that the night sky has eighty-eight constellations, but we only see about three-quarters of these in the Golden State? Some constellations, like the fabled Southern Cross, are only visible in southern latitudes. "When you see the Southern Cross for the first time," Crosby, Stills & Nash croon in their 1982 rock classic, "you understand now why you came this way." Likewise, folks living in the Southern Hemisphere have to come up our way to see the famous northern constellations, such as Ursa Minor, also known as the Little Bear or the Little Dipper. By the way, in our Northern Hemisphere we have the Northern Cross that's actually part of the constellation Cygnus the Swan. I've seen both the Southern and Northern Crosses and, in my opinion, our Northern Cross is just as nice and certainly larger!

A stargazer and his telescope (Photograph © Denny Long)

One of the coolest features of this book is the appendix of star maps. There are twelve—one for each month of the year—but here's the best part: They are set for Southern California skies! So you can take this book out in your back yard, flip to the correct month star map, look up at the sky, and there is a perfect map of your night sky!

Let me give you an example. Here in Southern California, we see Polaris—the bright star at the tip of the Little Bear's tail—about a third of the way between the northern horizon and the overhead zenith. That's because Polaris is almost directly above the North Pole. Wherever you are in the Northern Hemisphere,Polaris's height in the night sky just about matches your latitude. In Los Angeles (34° latitude), Polaris is 34 degrees above the horizon. If we lived up in Anchorage, Alaska (61° latitude), Polaris would be higher in our sky. If we lived down in Miami, Florida (26° latitude), Polaris would be lower. The farther south you go, the lower Polaris gets. If we lived at the Equator (0° latitude), Polaris would just barely rise above our horizon. If we lived in the Southern Hemisphere, we'd never see Polaris at all.

Here in Southern California, Polaris faithfully pops out every clear night of the year, about a third of the way up in the northern sky. And that's exactly where it shows up on the star maps in this book!

The star maps are also realistic. Unless you're in a mall parking lot, you should be able to see much, if not all, that's on the star maps. In my time, I've seen too many star maps supposedly designed for beginners that are way too complicated. Many shove in every star and minor constellation, and when you haul the maps outside and try to relate them to what you actually see, you wind up disappointed. If you're lucky enough to stargaze in the deep countryside, then, sure, the maps will make more sense, but only after many nights of practice, and that's provided you have night after night of clear skies. The reality is that most of us deal with at least some city lights, so I've designed my maps with that in mind. I don't want to undersell the stars, but I don't want to oversell either. Once you get into heavy-duty stargazing, go ahead and use more complicated star maps, but if you're new to all this, I think you'll like what's in this book.

I've been fascinated by the night sky all my life. I think we all are to varying degrees, even if you don't know what you're looking at when you gaze into the stars. You can gain a great deal of inner peace reclining on a lawn chair at twilight, watching the stars pop out, especially if you're wearing enough protection to keep mosquitoes away. Stay out even longer as that purplish darkness builds from the east and you're in for a treat. If you have a pair of binoculars handy, the experience will be even better.

I've been interested in astronomy and space travel since the 1960s space race between the U.S. and the U.S.S.R. During high school, I became a full-fledged amateur astronomer. I attended "moon watches" at a local nature center, and I'll never forget the first time I saw the moon through the instructor's large reflector telescope. I was hooked! With some help, I built my first telescopes and even ground the main mirror for one. I also took a beginning astronomy course in college, which served only to increase my passion for the night sky. I started teaching the basics of stargazing after college. These days, I still teach amateur astronomy and maintain a website, www.lynchandthestars.com, devoted to the heavens. In over thirty years, I have seen a lot of starry nights, but to this day (or night), I always see something new when I catch the latest show in the great celestial theater.

I love looking at the stars in the summer but I recommend it any time of year. Winter stargazing is more challenging but that's when the air is clearest and the constellations are the best. It's great to stargaze with friends and family, especially kids, but you should try it alone on occasion. I guarantee it will be good for your soul.

I don't try to solve the mysteries of the universe in this book. I'm not that smart and I don't want to kill too many trees. My goal is to teach you how to make the stars your old friends. I want you to be excited about looking into the night sky and making sense of it. I want you to be as turned on by astronomy as I am and so I've made this book as user friendly as possible. The great philosopher Socrates said, "Education is the kindling of a flame, not the filling of a vessel."

This book also tells the legends and mythology of the constellations. Through years of putting on stargazing shows, I have grown to love telling stories about how constellations got into the sky. Not all the celestial soap operas are exactly family friendly. Since this is a family book, they've been cleaned up, but not too squeaky clean. I don't want to take away all the fun! In my research, I've found that there can be dozens of stories for any given constellation. You have to remember that these stories are ancient and were mostly spread by word of mouth around campfires. You know what can happen. Stories get changed, even mutilated, along the way.

Constellation stories also vary by culture. Babylonians, Egyptians, Arabs, Chinese, and Native Americans, just to name a few, each have their own mythology of the stars. Most of my constellation lore is based on classic Greek and Roman mythology, though I have to warn mythology purists that I've thrown in a few modern elements. I figure that's okay if it helps you remember and repeat the stories and, besides, is there really such a thing as correct mythology?

I'll also explain the how's and why's of the changing night sky. Other than Polaris and a few northern constellations that rotate around the "North Star," we don't see the same stars every night and, while that makes stargazing more complicated, it also makes it more fun to keep up with the changes from hour to hour, night to night, and season to season. The earth's constant rotation on its axis and its annual orbit around the sun are both responsible for our changing vista of stars. At the same time, the moon and the planets take individual courses through the celestial sea. Comets, auroras, meteors, eclipses, and manmade satellites also come and go among the background of stars. The heavens are alive!

Whether you live in a neighborhood in Beverly Hills or in the desert around Death Valley, whether you're a beginning stargazer, an established amateur, or even a professional astronomer, I know you'll learn something from this book. More importantly, I want you to enjoy it and use it as your "go to" book for stargazing. Lay back and enjoy the greatest show out of this world.

This faint nebula, M78, is part of the famous Orion Nebula complex. (Photograph © Robert Gendler)

1

QUANTITATIVE MEDITATION

Before you get too far into this book, I want you to lie under the stars the next chance you get. A clearing in the woods, an open prairie, a quiet shore, or a backyard in the city are great places to begin your love affair with the stars. If it's winter, dress warm and haul along some blankets. When you stargaze, think about the facts and figures I'm about to lay on you. I call this quantitative meditation!

Milky Way (Photograph © Thomas Matheson)

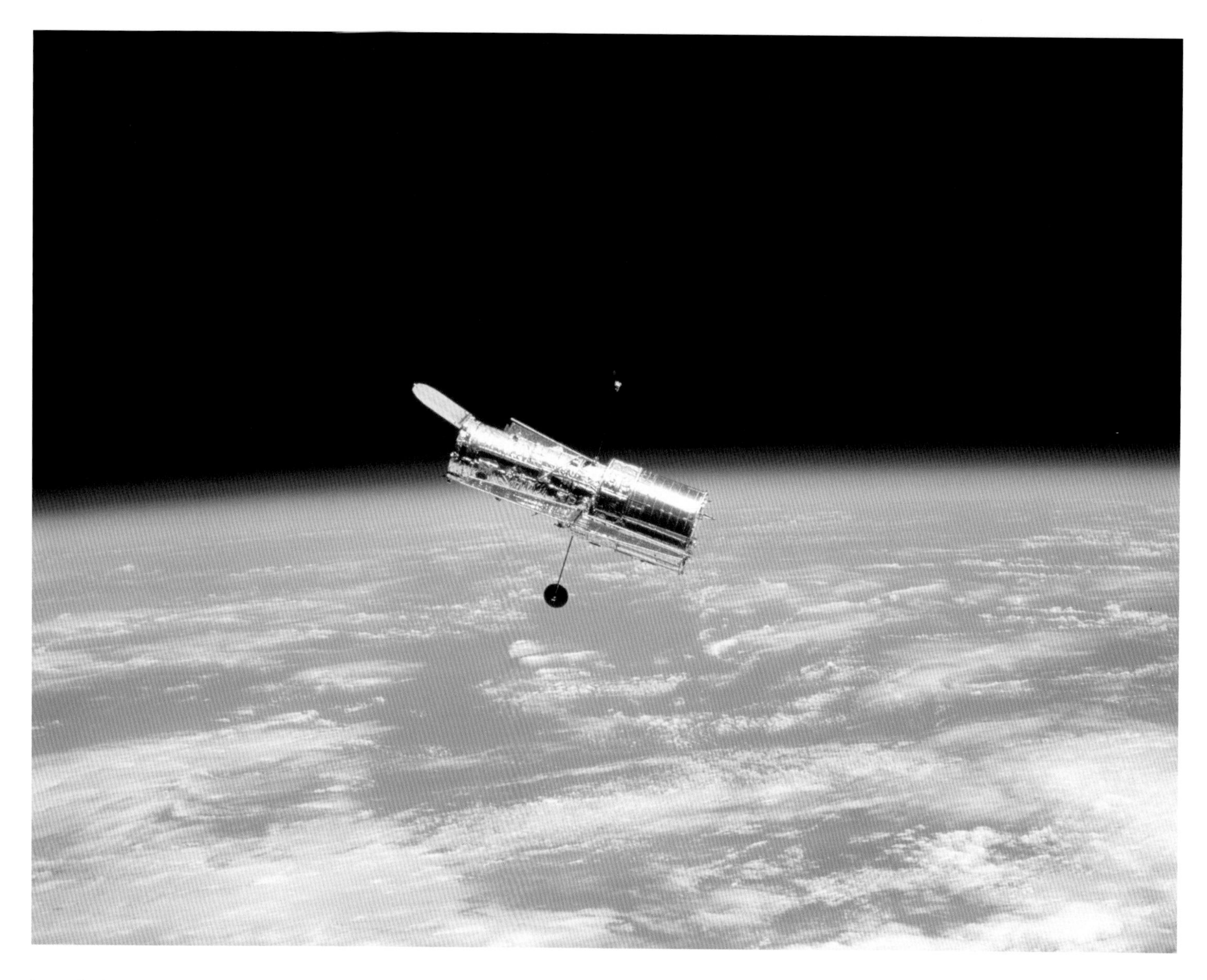

Launched in 1990, the Hubble Space Telescope has taken thousands of photographs in space. The pictures are wonderful, but don't be surprised when the view through your telescope doesn't match the intensity. Just remember that most Hubble pictures are colorized—and have a somewhat closer view of our universe! (Courtesy of NASA)

Let's start with the number of stars in the sky. Fifth-grade textbooks in my generation claimed that with the naked eye you could see about 3,000 stars in the night sky. Nowadays, unless you're in the boonies, you can't see anywhere near that many. Don't try to count stars or I guarantee you will fall asleep. Just lie back, relax your eyes, and you'll be surprised by how many stars you see. They pop out all over the sky. I call this "visually deep tracking." Believe me, this works, especially in the winter when the air is drier and clearer than at other times of the year.

The largest telescopes in the world—and those floating above this world, like the Hubble Telescope—can see many, many more stars than the naked eye. Some astronomers have estimated that our home galaxy alone, the Milky Way, contains up to a trillion stars, all revolving around a giant black hole at the galactic center. Our own star, the sun, nearly a million miles in diameter and 93 million miles away from the earth, takes over 225 million years to make just one orbit around that black hole in downtown Milky Way. As the sun obediently orbits, it drags along the earth and the rest of our solar system at over 130 miles per second. Speaking of the sun, it puts out the same amount of light as a trillion-trillion 100-watt light bulbs and has a surface temperature over 10,000°F. The sun is also 300,000 times more massive than the earth. While that size seems hard to imagine, more than half the stars in our night sky are bigger and brighter than the sun!

What about the distances to those stars? That 93-million-mile trek to the sun is celestial chicken feed compared to the distance to the rest of the stars. The next closest star to the earth is Proxima Centauri, over 25 trillion miles from your backyard.

To save us from the madness and the waste of printing endless zeros or dealing with cumbersome scientific notation, light years are used to express stellar distances. This measurement is based on the speed of light, a breakneck 186,300 miles a second. So how fast is that? If you took an airplane on thirty roundtrips between Miami, Florida, and Seattle, Washington, in less than a second, you would be traveling at the speed of light. A light year is simply the distance a beam of light would travel in a year's time. Just take 186,300 miles a second and multiply it by the number of seconds in a year, which is about 31.5 million, and you come up with a whopping 5.8 trillion miles for one light year.

The average distance to the stars you can see with your naked eye is about 100 light years. With 1 light year equaling close to 6 trillion miles, you're seeing starlight that has traveled an average of 600 trillion miles. Not only are you looking at objects that are astonishingly far away, but you're also looking back in time. If a star is 100 light years distant, then, by definition, it has taken an entire century for the light to reach your eyes. For example, Alkaid, the star at the end of Big Dipper's handle, is about 101 light years away. The light you see from Alkaid tonight left that star over 100 years ago.

While most of the bright stars are 100 light years away, give or take some, many stars visible to the naked eye are a lot farther. A good example is Deneb, one of the brightest stars in the summer sky and the brightest star in the constellation Cygnus the Swan. Deneb is over 3,200 light years away. That means the light we see from Deneb left the star before 1000 B.C. Deneb is believed to be more than 220 million miles in diameter and kicks out 300,000 times more light than our sun.

I know this quantitative meditation has more numbers than a typical income tax form. I don't want your head to explode as you try to relax under the stars, but let me give you a few more figures. If you could jump into a magical spaceship and fling yourself past the speed of light, which Einstein said was impossible, you could get out and away from our home galaxy. In your rearview mirror you would see this group of a trillion stars arranged in the shape of a giant CD broken up into spiral arms with a large hump in the middle. The Milky Way is a little more than 100,000 light years across and about 10,000 light years thick. The hump in the center of our galaxy is estimated to be 30,000 light years thick with a giant black hole in the middle. Just over eighty years ago, astronomers believed the Milky Way was all there was to our universe. But in the mid 1920s, the famous astronomer Edwin Hubble and his staff showed that other galaxies exist beyond the Milky Way.

The constellation Andromeda contains

MILKY WAY WITH EARTH AND SUN

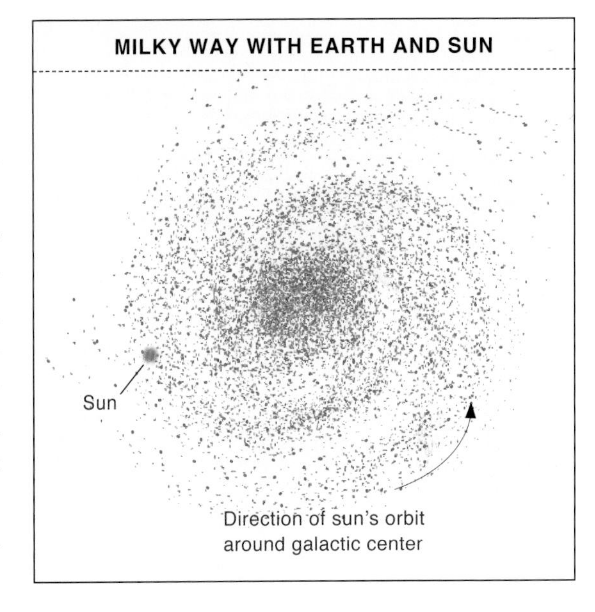

Like the Pinwheel Galaxy, M101, pictured here, the Milky Way is a spiral shaped galaxy. (Photograph © Robert Gendler)

ABOVE: Andromeda Galaxy, M31, in the constellation Andromeda (Photograph © Thomas Matheson)

FACING PAGE: These distant galaxies, the first that emerged after the Big Bang, represent our deepest view of the universe. Astronomers unveiled this picture, called the Hubble Ultra Deep Field, in March 2004. (Courtesy of NASA, ESA, S. Beckwith (STScI) and the HUDF Team)

a misty patch of light that astronomers, prior to Hubble's discovery, thought was a cloud of gas and dust, or nebula. By observing a certain kind of variable star within this luminous patch, Hubble concluded that the Andromeda "nebula" was actually an entirely separate galaxy. Consider it a next-door neighbor to the Milky Way, more than 2.5 million light years away! If it's dark enough where you live, you may see the Andromeda Galaxy with the naked eye, especially during autumn and early winter. Andromeda is the most distant object that can be seen with the unaided eye. The Apollo spacecraft that traveled to the moon in the late 1960s and early '70s made the lunar voyage in about three days. Keep in mind that the moon is only 222,000 to 252,000 miles from the earth at any given phase in its orbit. At the rate the Apollo traveled, it would take around 450 billion years for it to reach the Andromeda Galaxy!

By the way, we are traveling to the Andromeda Galaxy right now. Astronomers estimate that the Milky Way is headed toward our galactic neighbor at a rate of fifty miles a second. Circle your calendars for 5 billion years from now, when the Milky Way and Andromeda Galaxies will merge into one.

After Hubble's discovery of the Andromeda Galaxy, he and his followers found many more galaxies. They estimated that over 100 billion galaxies are accessible to earthbound and orbiting telescopes, spread across a universe that may stretch over 13 billion light years.

In my stargazing classes, I like to scale down sizes and distances for perspective. For example, if you shrink the sun's diameter from 864,000 miles to about the size of a period on this page, then on that same scale the next closest star, Proxima Centauri, is still more than 5 miles away! On this same scale, the diameter of the Milky Way is still more than 120,000 miles and the distance to the Andromeda Galaxy is 3 million miles.

Okay, you're probably "figured-out" by now. Some would say all these facts and figures take away from the joy of gazing into the night sky, but I think they enhance it . . . by light years!

2

WHAT'S A STAR?

What are those heavenly points of light that decorate our night sky? I could try to give you a highly technical and academic definition of a star, but let's keep it simple. Stars are basically big glowing balls of gas. Each one is an independent, self-contained giant nuclear plant. Our sun, an average-sized star, is just a ball of hydrogen gas with an 864,000-mile girth and a surface temperature over 10,000°F.

How do stars shine? When I pose that question to my beginning stargazing classes, some of the answers I hear are "friction," "reflection," "refraction," "heat," "the sun," and "the leftover glow from the Big Bang." I once watched a five-year-old girl with pigtails get a big smile and say, "Stars are the lights from heaven leaking into outer space." In all my years teaching, that's my favorite response. The most common answer I get is "gas burning up." That's close but no celestial cigar!

The sun (Photograph © Jack Newton)

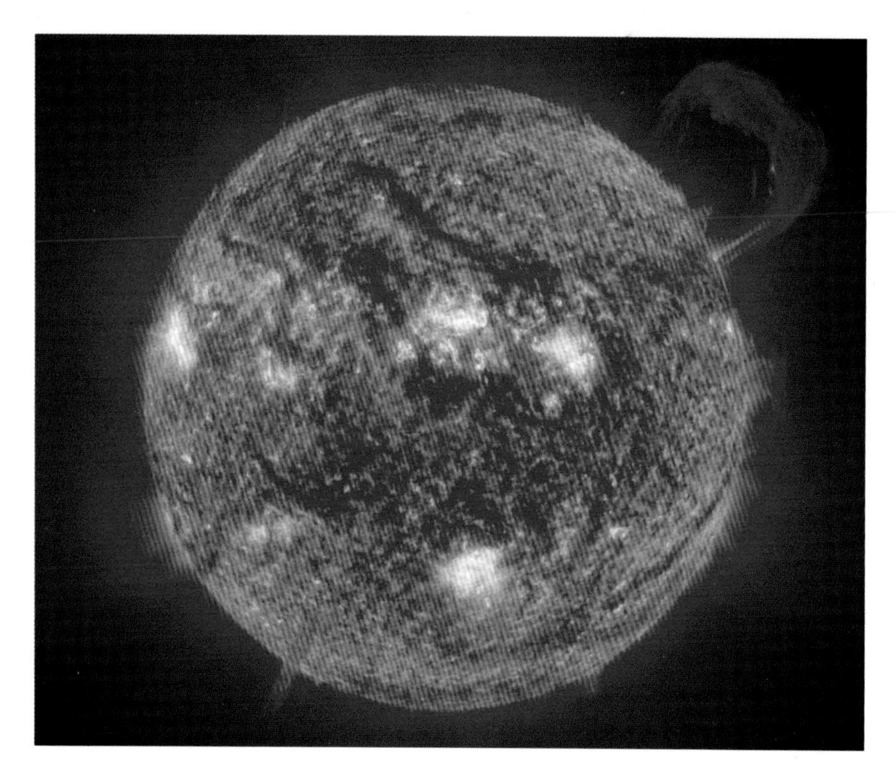

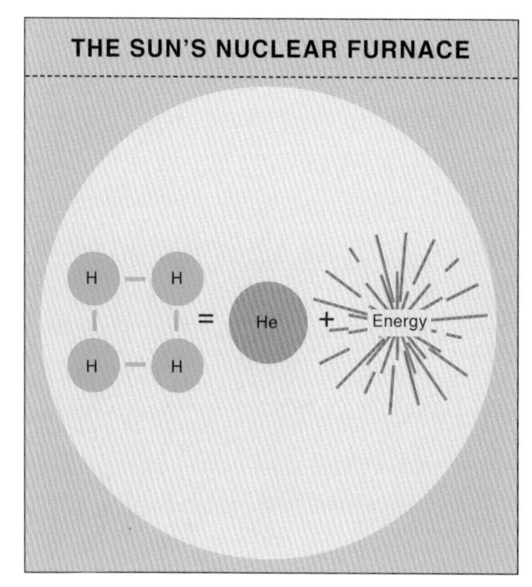

TOP: The sun (Courtesy SOHO/EIT consortium)

So how *do* stars shine? How does the stellar furnace operate? It's a simple one-word answer: gravity. These gigantic balls of hydrogen gas are being held together by their own gravity, and as they squeeze inward, the interior temperature skyrockets. The force of the gravity pulling in on a star is proportional to its mass. Now you may think these balls of gas wouldn't be all that massive, but believe me, they are. In fact, the mass of our sun, in kilograms, is 2, with 30 zeros after it (that's 2,000,000,000,000,000,000,000,000,000,000)! If you're not that metric friendly, 1 kilogram equals about 2.2 pounds. Stars are heavy! The resulting gargantuan gravitational force relentlessly attacks the star from all sides. At our sun's core, this gravitational pressure exceeds over 500 billion pounds per square inch, which drives up the core temperature by millions and millions of degrees. You can't exactly stick a meat thermometer inside the sun but it's estimated that the core temperature is 27 million°F.

The core of every star is extremely hot. How hot, you ask? Hot enough to create nuclear fusion. Nuclear fusion is a complicated process, but essentially it comes down to this: hydrogen atoms fuse together to form heavier helium atoms and, in the process, a tiny amount of hydrogen converts into energy in the form of light, heat, gamma rays, x-rays, and more. This radiation slowly makes its way from the star's core to its surface. This journey can take thousands of years, but once it makes it to the surface, the radiation then travels in all directions at the speed of light.

To recap, hydrogen is the fuel of a star, helium is the ash, and energy is a byproduct. In just one second, our sun converts almost 700 million tons of hydrogen to helium and energy. In spite of that incredible rate of consumption, our sun won't run out of hydrogen for another 5 billion years!

THE BIRTH OF STARS

Where do stars come from? They have to start somewhere, and the answer is gas. All stars form out of nebulae, huge clouds of hydrogen gas and dust. Everywhere in our galaxy and across our universe of galaxies are hydrogen clouds. Hydrogen is the most basic of all elements, and shortly after the Big Bang (the theorized beginning of the universe, about 13.5 billion years ago) hydrogen was pretty much all there was. Ever since that time, stars have been born out of these nebulae. Nebulae also come from the residue of exploded stars.

Gravity waves—from exploding stars, passing stars, or galactic tides—cause pockets of nebulae to condense and form "proto-stars." Hundreds of these proto-stars can form at the same time in a nebula, along with developing solar systems like our sun and its orbiting planets. Over millions of years, these proto-stars gain material due to their growing gravitational force, and eventually they acquire enough mass and gravitational pressure to fire up nuclear fusion. Then stars are born and fall into what astronomers call hydrostatic equilibrium. Their own internal pressure pushing out keeps the stars intact and balances the gravitational pressure pulling in.

During the first few hundreds of millions of years, a cluster of young stars born out of the same nebula hangs out together in one big happy family,

Called the Fingers of Creation, this photograph captures newborn stars emerging from gaseous pillars in the Eagle Nebula, M16, in the constellation Sagittarius. (Courtesy Jeff Hester and Paul Scowen, Arizona State University, and NASA)

The Dumbbell Nebula, M27, is a planetary nebula in the constellation Cygnus. (Photograph © Robert Gendler)

bound not by love but by gravitational attraction. Over the eons, as these stellar families journey around the center of their home galaxy, gravity from passing stars breaks the clusters apart until most of the members are on their own, though some of the stars from the birthing cluster stick together in groups of two, three, or more. When you scan the night sky with binoculars or a telescope and see a cluster of stars, chances are you're seeing baby stars.

THE DEATH OF SMALLER STARS

The lifespan of a star is dependent on one thing and one thing only: mass. If you're a star and you want to live a long time and have a relatively quiet death, be a low-mass star. If you want a short life and you want to go out with a bang, then you should be a massive star—a star at least eight times more massive than our sun.

Low-mass stars like our sun can live over 10 billion years. Even though our sun consumes 700 million tons of hydrogen every second, big-time massive stars use their huge hydrogen supply at much higher rates. Some of the giant stars, like Deneb in the constellation Cygnus, are only around for a few billion years.

Low-mass stars get fat before they die and flicker out. In the case of our sun, it will run out of hydrogen

TOP: The Beehive Cluster, M44, also known as the Praesepe, is an open cluster of young stars. (Photograph © Russell Croman)

in about 5 billion years. In the sun's core, helium builds up as the hydrogen dwindles. As nuclear fusion dies out, internal pressure will decrease. The built-up helium in the core will start to contract because of the never-ending gravitational pressure, which will cause the helium core to increase in temperature. Some of the heat will escape the core and reach cooler hydrogen layers. In time, the temperature will rise enough in these outer layers to fire up nuclear fusion, which in turn will cause the sun to bloat. The result is called a red giant star. When that happens, our sun's diameter will expand from 864,000 miles to about 75 million miles and the sun will swallow its closest planet, Mercury.

As time goes on, the heat in the core of the new red giant sun will build up even more until helium fuses into heavier elements like carbon and oxygen. The energy produced with this kind of fusion will drive more heat to the outer hydrogen layers, causing that hydrogen to fuse to helium. That will produces even more heat, and the red giant will respond by swelling out even more, becoming a super red giant that swallows the next nearest planet, Venus. The outer edge of the swollen, dying sun will almost touch the earth and, needless to say, we'll be toast! Even though the sun's surface temperature will be cooler, about 3,000 to 4,000°F, it will be right on top of us. Actually, life as we know it will be long gone well before then. Even now, our sun's temperature is slowly increasing, and many astronomers predict that in about a half a billion years, the sun will be so hot that the oceans on Earth will completely evaporate.

Toward the end of our sun's red-giant phase, excess energy "burps" in the outer layers will cause large clouds of gas to blow off and form shells of gas around what's left of the star. We call these planetary nebulae and, with a moderate telescope, you can see plenty of them in our night skies, surrounding stars that are biting the celestial dust. Some, like the beautiful Ring Nebula and the Dumbbell Nebula, can be seen high in the summer skies, even with small telescopes.

After 1 or 2 billion years as a red giant, stars that have either consumed all their hydrogen in fusion or have blown it off begin to shrink into white dwarfs. With no more nuclear reactions within the star to hold it together, gravity collapses the corpse of the once proud star. In our sun's case, it's believed that half the sun's original mass will be squished into a ball the size of the earth, about 8,000 to 10,000 miles in diameter. A white dwarf is so dense that 1 tablespoon may weigh as much as 1 to 2 tons in Earth weight. That's the weight of most SUVs. Imagine a tablespoon holding up a Chevy Envoy.

A DEVELOPING RED GIANT

Heat
Heat
He
H fusion

White dwarfs are "retired" stars that initially emit some light but eventually fade and become dark dwarfs, which is not the same as a black hole. That fate awaits the super massive stars.

THE DEATH OF BIG, BIG STARS

Massive stars are short timers. One to 2 billion years of normal life is usually all they get before they squander their hydrogen fuel. Just like low-mass stars, they turn into red giants, but these red giants are much larger than their lower-mass brethren. One example is Betelgeuse, the second brightest star in the constellation Orion the Hunter. Betelgeuse is a pulsating variable star, which at its greatest girth

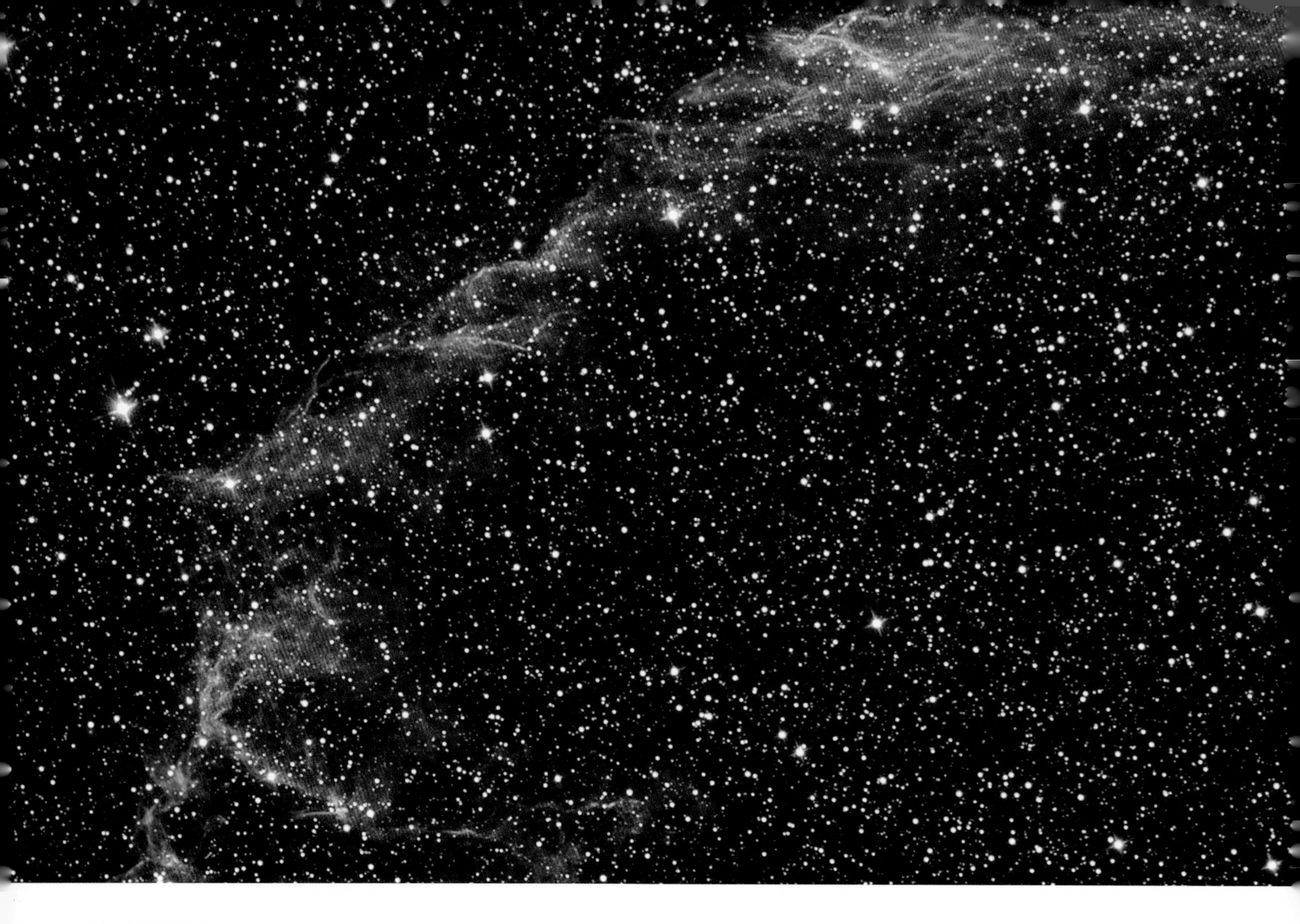

The Veil Nebula is a supernova nebula in the constellation Cygnus. (Photograph © Rick Krejci)

is over a billion miles in diameter. If you were to put Betelgeuse in our solar system in place of the sun, its outer edge would reach nearly to Jupiter. Mercury, Venus, Mars, and Earth would all be inside Betelgeuse!

The same process that turns less massive stars into red giants also applies to these big guys. Helium atoms inside the stellar core fuse in stages to heavier carbon and oxygen, releasing energy along the way, but because these stars are more massive and have greater gravitational pressure, nuclear fusion continues. So much heat builds up in the core that heavier and heavier elements are created until finally iron is produced. That's the end of the line, because it's not physically possible for the star to fuse iron atoms into heavier elements. The iron core then compresses due to the unrelenting gravity.

The core collapses into such a compressed ball that nothing but neutrons are left, but it is not a steady shrinkage. The core rebounds, bouncing back as it compresses. It's like a person jumping on a trampoline. While this activity is going on, stellar material around the original iron core falls into the void. That's when all nuclear you-know-what breaks loose. The material falling inward runs into the atoms of the iron core bouncing off the trampoline and you have an explosion beyond your wildest dreams. It's similar to a hitter taking a mighty swing at a 90-mph fastball and making perfect contact. The ball explodes off the bat and it's out of the park.

The super red giant explodes into a supernova, flinging out material at incredible speeds of 10,000 miles a second! For a few months, the exploded star can emit 10 billion times more light than our sun and outshine any star in our night sky. In some cases, it can be bright enough to be seen during the day! In time, these supernovae fade. One example of a faded supernova is the Crab Nebula, located in the constellation Taurus the Bull. This phenomenon doesn't occur that often. A galaxy of over 500 billion stars might experience only two supernovae in a century.

Not only are supernovae great celestial fireworks, they're vital to the development of our universe, because that explosion is the only way heavy elements like gold, silver, uranium, and many others are created. These elements are uniquely produced by nuclear synthesis during a stellar explosion. No other process makes this stuff. These elements become the building blocks for new stars, new planets, and possibly new life. The iron in your blood and the calcium in your bones are the result of some supernova somewhere back in time. Exploding stars are part of our circle of life!

What's left of an exploded star after a supernova can be one of two bizarre objects: either a rapidly rotating neutron star or a black hole, depending on the mass of the remaining core. If the mass of the core is about twice the mass of the sun, it's considered a neutron star or a pulsar, a star that's as dense as the nucleus of an atom. Neutron stars are only 10 to 15 miles in diameter, give or take a few, and they are so dense that 1 tablespoon weighs over a billion Earth tons! If you were to squish a skyscraper to the density of a neutron star, the building would be approximately the size of a quarter. Neutrons stars also rotate at speeds of over 600 times a second. As they rotate, they emit pulsating radio waves and some visible light waves. When these radio pulses were first discovered in the 1960s, some folks thought we were getting communications from aliens. But astronomers quickly discovered the true nature of these unusual pulses.

If the core left behind by a supernova is more than two solar masses, caused by a star possibly over fifteen times the sun's mass, you have a black hole on your cosmic hands. These enormous red giants collapse rapidly, possibly in a matter of hours, to objects so small and dense, and with so much gravity, that not even light can escape. What goes into a black hole never comes out! Some would say it's like sinking tax dollars into certain government functions. Seriously, black holes are probably the most bizarre objects in our universe. The escape velocity of a black hole—that is, the speed you would have to reach to escape the gravity pull of a black hole—is more that the speed of light, so light is trapped forever in a black hole. Obviously, nobody can directly see a black hole because there's nothing to see. However, astronomers have found evidence of their existence.

The first candidate for a black hole was detected in the constellation Cygnus the Swan. An x-ray telescope detected a huge energy source in the form of x-ray radiation, and astronomers designated it Cygnus X-1. The theory is that x-rays are being produced by a black hole sucking gas off a neighboring star. The gas is flying so quickly into the black hole that x-rays are being produced. Since then, many other candidates for black holes have been found.

Astronomers believe that super-massive black holes exist all over the universe and particularly gigantic ones are found in the centers of galaxies. The center of the Milky Way may be a black hole with a very small volume but weighing around 2.5 million solar masses! Our next-door neighbor, the Andromeda galaxy, may have a 45-million-solar-mass black hole. Some of the super giant galaxies of our known universe have possibly 3-billion-plus-solar-mass black holes in their celestial downtowns.

COLOR IN THE STARS

Stars are not just white lights. Although it won't leap out at you like the NBC peacock, you can see tinges of color in the night sky with a little extra staring. Viewed through binoculars or a small telescope, stars really show their true colors.

The subtle coloring of stars ranges from bluish white to reddish white, and in-between are hints of yellow and orange. By observing a star's color, you can get a good idea about its nature, especially temperature. It's simple:

Quintuplet Cluster (Courtesy of NASA, Don Figer, STScI)

the bluer the star, the hotter its surface; the redder the star, the cooler its surface. It's the same with any fire. When you stare into a campfire, the hottest flames are blue, yellow ones are a little cooler, and the coolest flames are red.

One of the bluest stars in our sky, Spica, has a surface temperature of over 36,000°F. If, on the other hand, you were to toe-test the super red giant star Betelgeuse, your toe would feel about 5,800°F. It's relatively cooler, but you'd still be better off walking on hot coals! Our sun, with a surface temperature of just over 10,000°F, is considered a yellowish star. However, any astronaut who's ever been in space will tell you the sun is blinding white.

The color of stars also reveals other characteristics. Most reddish stars like Arcturus are near the end of their lives. They're red giants that are bloating and depleting the last of their hydrogen fuel. Depending on how massive they are, they'll eventually shrink into white dwarfs or blow up in supernovae. Bluish stars like Spica are so giant to begin with that most have a short lifespan. They gobble up hydrogen fuel at incredible rates, expand into super red giants, and eventually explode in supernovae.

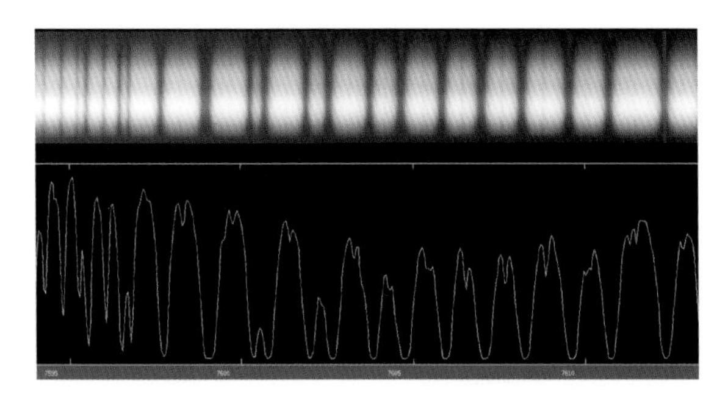

This stellar spectrum is from a red super giant in Canis Major. The displayed spectrum covers a wavelength range deep in the red beyond what the human eye can detect. (Courtesy of Gerald T. Ruch Jr., University of Minnesota)

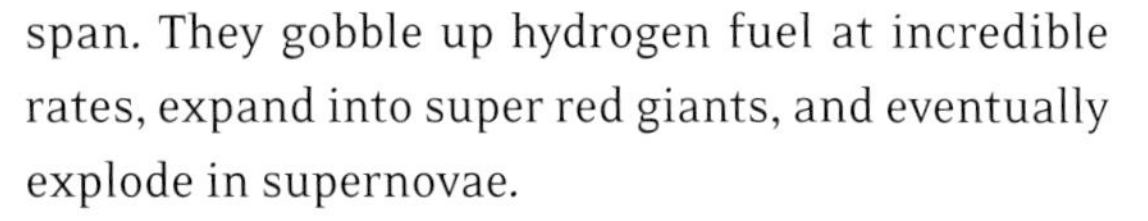

BREAKING THE STELLAR CODE

With few exceptions, when you look at any individual star through a telescope—from the smallest backyard scopes to the Hubble Space Telescope—all you see is a single point of light. Some of the giant stars like Betelgeuse have been resolved, or clearly seen, as a disk, and in the future more and more stars will be resolved, as telescopes and techniques become more sophisticated.

Even when that happens, most of what we know about individual stars will still come from the spectrum of a star. From the celestial "fingerprint" emanating from that single point of light, light years away, you can determine what elements are found in and around that star. You can determine its temperature, its movement toward or away from the earth, and whether or not it has a companion star or stars, or perhaps a planet system circling it. You can also indirectly determine the age of the star, its distance from Earth, and its rotation rate, all from that single point of light run through a spectroscope.

Starlight is split into its color components, or wavelengths, the same way you strain sunlight through a glass prism to get a rainbow. You can use natural prisms like the spray of water from a garden hose. Adjust the nozzle to a light spray and hold toward the sunlight at just the right angle, and the droplets of water will create a mini rainbow or spectrum of our closest star's light. Big-time rainbows are created the same way, only on a larger scale. Water drops in clouds and rain are at just the right angle after a spring rainfall to refract sunlight into its full spectrum of colors.

Spectrographs attached to telescopes create a much more detailed spectrum. Dark lines showing up at various wavelengths on the spectrum act as the fingerprints of various elements present in and around the star. We know the dark lines signify certain elements, such as hydrogen or helium, because of laboratory observations of the same elements here on Earth.

Where the dark lines fall on the spectrum lets astronomers know whether the star is moving away from us or toward us. In the visible spectrum, blue light comprises shorter wavelengths and red light comprises longer ones. In the case of receding stars, the wavelengths stretch out and appear longer, placing the dark lines on the red end of the spectrum; for approaching stars, the wavelengths compress, making them appear shorter and placing them at the blue end. This change in wavelength is called the Doppler effect.

We experience the same phenomena with the sound waves from an ambulance siren. The sound changes from a higher pitch to a lower pitch as the siren passes us. When the vehicle approaches, the sound waves compress, giving them a higher pitch. When the vehicle has passed us, the sound waves stretch out, giving them a lower pitch.

After thousands of stellar spectrums were

gathered, astronomers discovered that many were similar. They grouped stars in a set of spectral types according to which lines were the darkest on its spectrum. It turned out that stars with similar spectral types had similar surface temperatures and so the spectral types were arranged by temperature from hottest to coolest. The spectral types are O B A F G K M R N and the best way to remember them is with the archaic saying "Oh, be a fine girl; kiss me right now!" Obviously, back when spectral types were established, most astronomers were men. As you can see on the chart, O stars are the hottest and N stars are the coolest. A spectral type can then be subdivided in groups from one to ten, according to the star's temperature. For example, an O1 star is hotter than an O7 star.

In the early 1900s, astronomers Ejnar Hertzsprung of Holland and Henry Norris Russell of the United States made an amazing discovery. They studied the spectrums of thousands of stars and found a definite relationship between spectral type and luminosity, which is the amount of light a star produces. In fact, they found that most stars could be put on a graph and fit along a nice curve. They then developed the famous Hertzsprung-Russell diagram, a key step in understanding stars and their evolution. Different types of stars are on different parts of the curve and some, like dying red giants, are off the curve. The bottom line is that by just studying a star's spectrum, or fingerprint, there's much information to glean.

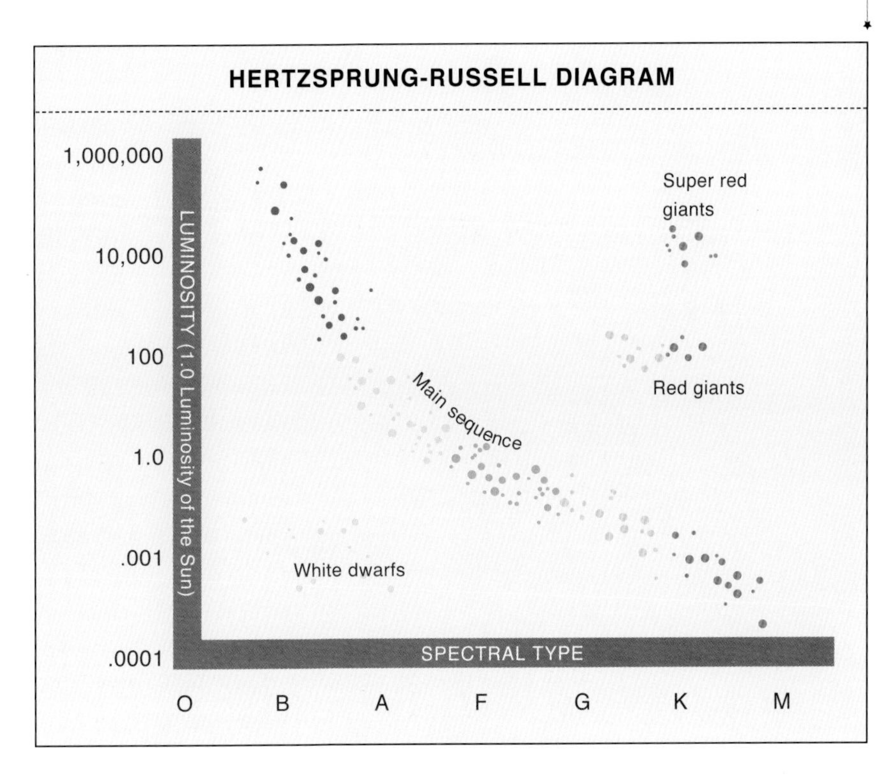

DOUBLE STARS, TRIPLE STARS, AND MORE

When you wish upon a star, chances are you're pinning your dreams on more than just one star. More than half the stars in our night sky are actually double or triple stars revolving around each other. Astronomers have also found quadruple, quintuple, and sextuple stars that appear as a single star to the naked eye. Our home star, sitting all by its lonesome, is not the norm.

While Castor, the brightest star in the constellation Gemini, appears to be a single bright star, looks are deceiving! Gazing through a moderate telescope reveals two separate stars. Industrial-strength telescopes reveal that Castor is actually made up of six stars revolving around each other. If you lived on a planet in that solar system, you would have six sunrises and six sunsets every day. Your calendar would be a mess and your climate chaos!

I think the best double star is Albireo in the constellation Cygnus the Swan, because it's colorful and easy to split. To the naked eye, Albireo looks like a single, moderately bright star at the head of the swan, but with a small telescope or a good pair of binoculars, you can see a lovely contrasting double star. One star has a gold hue and the other is a distinct blue. It's one of the gems of the sky!

VARIABLE STARS

Most stars, including our sun, are steady shiners. For thousands and thousands of years, our sun hasn't varied from its present brightness by much more than 1 percent. However, some blinkers out there, dubbed variable stars, change their brightness in a matter of hours, days, or hundreds of days. About ten or eleven can be seen with the naked eye.

Some variable stars fluctuate in brightness because of internal changes. They go from one nuclear fusion process to another, which changes their light output. Others vary because they are in a binary system, with one star eclipsing the other as they orbit each other.

One of the best variable stars is Mira in the faint constellation Cetus. In 332 days, Mira goes from a fairly bright third-magnitude star to a dim ninth-magnitude star that you can only see with a telescope. Another great variable star is Algol in the constellation Perseus the Hero; it doubles in brightness then fades in less than three days. One of the Greek myths has Perseus killing Medusa, the snake-haired monster who was so ugly that anyone who looked at her turned to stone. The constellation Perseus is portrayed as the conquering hero flying back from his mission, holding the severed head of the monster. Algol is the blinking eye of Medusa. Don't look too close. I don't want you turned to stone!

STELLAR PARALLAX METHOD

Parallax
Position of star in December
Position of star in June
Parallax angle
x
93 million miles
Sun
Position of Earth in June
Position of Earth in December

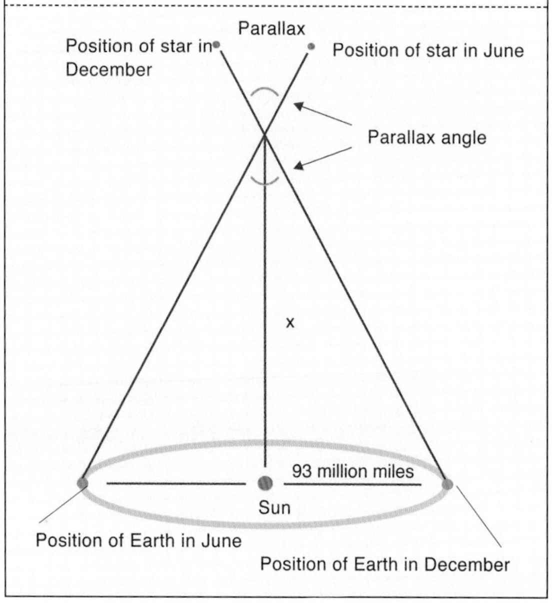

HOW FAR AWAY ARE THE STARS AND HOW DO WE KNOW?

I've rattled off distances to many different stars and galaxies. How do astronomers know how far away these stars are? Do they just point their telescopes at the sky, examine a few stars, then scratch their heads and say, "Because I have a PhD, I think that star is 100 light years away and that one over there is 5,000 light years away?" No, there's more to it than that, but it's not too complicated, at least not for the closer stars.

For stars about 500 to 600 light years from the earth, you use the stellar parallax method. You take a picture of a star when the earth is on one side of the sun and you take another picture six months later when the earth is on the other side of the sun. The star will appear to shift a tiny bit against the background of more distant stars. The apparent shift is called a parallax. It's like when you hold a pencil at arm's length and observe it with your left eye shut and then your right eye. The pencil will appear to shift position against objects on the far wall. In both the case of the pencil and the star, the object hasn't really moved, but the apparent shift can help you determine the distance of that pencil or distant star.

This process comes down to simple high-school trigonometry. Your next step is to use the apparent shift to come up with a parallax angle. Using the rule that says opposite angles are equal, you draw a triangle between the star and your two points of observation, with one leg of the triangle slicing through the sun. You take the parallax angle you so carefully measured and cut it in half, as shown on the diagram. Since you know the sun is 93 million miles away, and you know the measurement of one angle, it's simple "trig" from here. The distance to the star equals 93 million miles divided by the tangent of the parallax angle.

As simple as the math is, the practice of measuring that parallax angle is difficult and you're also making assumptions. You're assuming the background stars you used to measure the star's apparent shift are stationary. In reality, they may shift as well!

The stellar parallax method is also difficult to perform from the earth's surface because you put up with blurring atmosphere. That's why the Hipparcos satellite was launched in 1989 to measure the stellar parallax of hundreds of stars. Despite its success, the satellite's accuracy drops when it tries to measure larger stellar distances and their smaller parallax angles.

Stars beyond 500 light years require another method, which uses the famous Hertzsprung-Russell diagram. If you know the spectral type of a star, you can use the Hertzsprung-Russell diagram to deduce its luminosity. Once you know the luminosity, figuring out the distance is an easy math equation using the simple inverse-square law of light, which you'll learn about in the next chapter.

For really distant stars, yet another method of measurement is used. This was a huge discovery made early in the last century by Henrietta Leavitt at Harvard University. She studied thousands of variable stars and discovered that some are extremely

Cepheid variable stars shine in a distant spiral galaxy, NGC4603. (Courtesy of NASA/HST Key Project Team)

regular in their pulsation and extremely bright, shining 500 to 10,000 times the sun's luminosity. Leavitt found that these stars, called Cepheid variables, have a near-perfect relationship between their period of variation and their average luminosity—and once you have luminosity, it's simple math to calculate the star's distance from the earth. Because of their brightness, Cepheid variables could then be used as mile markers in deep space.

Edwin Hubble observed Cepheid variable stars in what was then known as the Andromeda Nebula to determine that Andromeda was not a gas cloud but a whole other galaxy. Since Hubble's time, billions of galaxies over 13 billion light years away have been found. The Cepheid variable is one of the handiest tools ever used by astronomers. What a great celestial yardstick!

HÆMISPHÆRI
GRAPHICUM
COELI
TIET
Andromeda
Pegasus
Scheat alferas
Delphinus
ARIES
PISCIS
AEQVINOC
Cetus Balæna
Baten Kaitos
Daneb Kaitos
CIRCVLVS
AQVARIVS
CAPRICORNVS
SAGITTARIVS
OCEANVS
ÆTHIOPICVS
Tropicus
Piscis Aust:
Phoenix
Grus
Geranos
Indus
Toucan
Pica Brasilica
TER RA AVS TRALIS IN COGNITA
Hydrus
Pavo
Corona Aust
Apis Indica
Chameleon

3

A SKY FULL OF CONSTELLATIONS

When experienced stargazers look into the night sky, they see lions, bears, hunters, a goose, a giant scorpion, a dolphin, a harp, and a lot of other animals, people, and things. These are the constellations. Allegedly, the stars make pictures in the sky, but the problem is that most constellations don't look like their namesakes. This isn't only because of city lights that wipe out all but the brightest stars. Even in the dark countryside, you're forced to put your imagination into overdrive! In this modern age, we still view constellations the same way cave people did. We draw imaginary lines between the stars to come up with images, just like dot-to-dot puzzles in a kindergartner's coloring book. Unlike the coloring book, though, the night sky doesn't number the stars. We have to decide what stars to connect with our mind's eye.

Antique constellation map

Cygnus the Swan and other constellations in the summer Milky Way (Photograph © Thomas Matheson)

Ever since ancient times, these pictures in the stars have commemorated a character or object. Local mythology dictated which constellations were which. Back then, constellations were much more important to a culture than they are now. Without shopping malls or streetlights, the skies were a lot darker and folks couldn't help but be overwhelmed by what they saw. Many famous constellation stories come from ancient desert civilizations that saw the stars night after night after night. They believed their god or gods provided these cosmic pictures.

Constellations were a great vehicle for storytellers to pass down cultural myths and legends from generation to generation. Back then—before the sensory overload of books, DVDs, or text messages—no one cared that constellations didn't quite match what was being portrayed. Around a campfire the seemingly never-changing patterns were handy storytelling tools. Constellations were also great for passing down lore because the star patterns never seemed to change in size or shape. In reality, the stars change positions relative to each other as they orbit at various speeds around the Milky Way, but not enough to radically change constellations for thousands of years. We still see basically the same constellations in the Space Age as were seen in the late Stone Age!

No one knows exactly when people started seeing these figures in the sky. Some artifacts and ancient texts that go back more than 5,000 years to ancient Sumeria, now present day Iraq, show that people recognized constellations. Scholars have concluded that this ancient civilization already had a lion, a bull, and a scorpion constellation. We see these in our skies today, in more or less the same form, as Leo the Lion, Taurus the Bull, and Scorpius the Scorpion.

Other civilizations, such as the Egyptians, Chinese, Babylonians, and Native Americans, developed their own constellation lore. The funny thing is that many different cultures, separated by vast distances, came up with similar interpretations for some of the same groupings of stars. For example, many separate cultures saw the Big Bear as the Big Bear. To this day, how and why this happened is still a mystery.

By the time the second century rolled around, the famous Greek astronomer Ptolemy had cataloged forty-eight constellations, some of which he borrowed from the Babylonians and other cultures. He published these constellations in his great work, the *Almagest.* The Greek lore of the constellations is colorful and many of the stories are well known to this day. You could call them the earliest soap operas, and they're not exactly rated G. They're based on Greek mythology involving Zeus and Hera, the king and queen of the gods, and their dysfunctional courtship and marriage. Later on, the Romans ripped off—or should we say borrowed—the Greek constellations and slapped on some of their familiar Latin names and mythology.

STAR NAMES

After the fall of the Roman Empire, the Arabs, who were great astronomers and stargazers, preserved records and wrote books on the forty-eight constellations and gave names to many of the brighter stars. These names usually referred to the position of a star in a certain constellation. For example, the bright star Rigel, in the constellation Orion the Hunter, translates to "foot." Betelgeuse, at the other corner of Orion, roughly translates into "armpit of the mighty one." Some stars have Greek and Latin names as well, but most traditional star names are Arabic.

In 1603, German astronomer Johann Bayer introduced a more astronomical method of naming stars. In his star atlas *Uranometria* Bayer used Greek letters to name and organize the stars, constellation by constellation. The brightest star in each constellation, based on naked-eye observation, was designated the first letter in the Greek alphabet, alpha (a). The second brightest star was beta (b), the third brightest was gamma (c), and so on. For example, the brightest star in Ursa Major is Alpha Ursa Majoris. Majoris is the possessive form of Major, so the name in English is Alpha of the Big Bear. The traditional name for this particular star is Dubhe (pronounced Dubby); other bright stars like Capella, Betelgeuse, and Vega have also maintained their traditional Arabic, Greek, or Latin names. However, the Bayer naming system is full of inaccuracies. In many constellations, the star listed as alpha is not the brightest. For example, Orion's alpha star is listed as Betelgeuse, but its brightest star is actually Rigel.

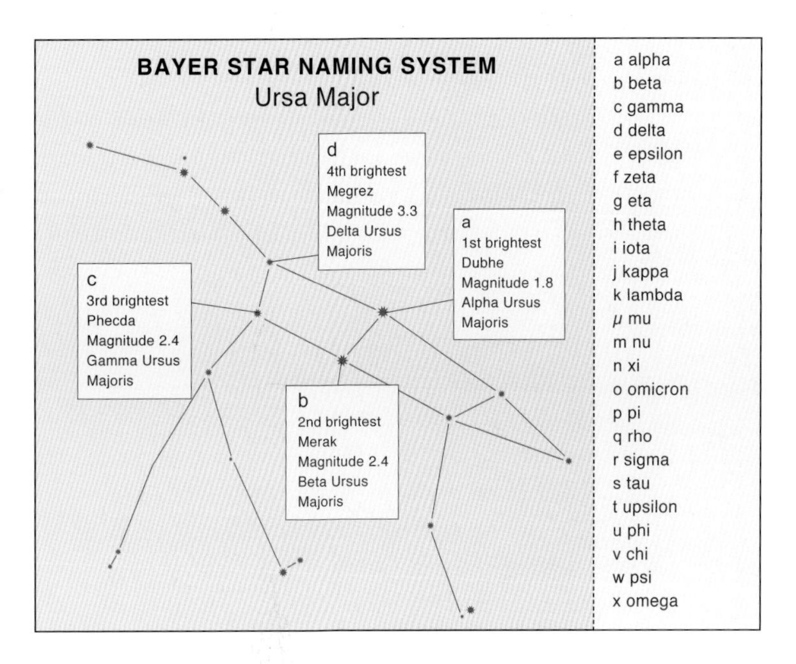

Another problem with the Bayer system is that there are only so many Greek letters. By the end of the seventeenth century, British astronomer John Flamsteed addressed this shortcoming by introducing numbers to each star not covered by a Greek letter. Flamsteed's numbers are applied in ascending order from west to east across a constellation. In this book's star maps and charts, you'll see traditional names and Bayer's Greek letters, but not Flamsteed numbers. In order to make the stars your old friends, it's good to keep things simple at first.

NEW CONSTELLATIONS!

In modern times, during exploration of the southern hemisphere, sailors and navigators returned with tales of constellations never seen in Europe or the Mideast. In the early 1600s, Johann Bayer, in addition to coming up with Greek letters to name the stars, named and cataloged twelve new constellations based on these observations. Even more constellations were added in later centuries, and constellation confusion set in. Multiple constellations were occupying the same patch of sky. Something had to be done! So in 1930, the International Astronomical Union came up with a list of eighty-eight constellations that are still recognized to this day. Every part of the sky is occupied by a single constellation. Their boundaries are sectioned off with parallel and perpendicular boundaries.

Not all constellations are created equal. Some are big and some are small, some bright and some dim, and some are in between. In states that lie roughly along the 45° latitude, we see about sixty-six constellations through the course of a year. The other twenty-two never get above our horizon and remain treasures of more southern latitudes. Of the visible sixty-six constellations, about forty-four can be seen at any time in the night sky. To be honest, only about half of those constellations are big and bright enough to be worth looking for, especially if you're new to stargazing. The rest are dim or tiny or both.

While you're watching the unfolding drama in the sky, keep in mind that constellations are perceived formations, not actual ones. While constellations appear to be two-dimensional, we are looking into three-dimensional space. You can't fire up an advanced version of the space shuttle and fly to Orion the Hunter, because its stars are at different distances from the Earth. Some are less than 100 light years away; some are thousands of light years distant.

THE BRIGHTNESS OF STARS

Some stars twinkle brightly in the night sky, while others fade into the background. On most star maps, the brightness of a star is represented by the size of the dot. The bigger the dot, the brighter the star. To get the most out of the charts and maps in this book, pay attention to this feature. It will make a big difference in your stargazing success.

There is a way to quantify the brightness or dimness of a star and it's called the apparent magnitude scale. It was developed back in the second century B.C. by the famous Greek astronomer Hipparchus. Hipparchus, without the aid of a telescope, divided stars into six groups. The screwy part of his method is that the brighter the star, the lower the number it was assigned, and the dimmer the star, the higher the number. I would think it would be the other way around, but who am I to question a great Greek astronomer? Hipparchus designated the brightest stars as first-magnitude stars and the dimmest as sixth-magnitude. The system is based mathematically on the fifth root of 100. In plain language, that means that the brightest stars in the sky are 100 times brighter than the dimmest stars. The difference in brightness between each magnitude is a factor of about 2.5. In other words, a first-magnitude star is 2.5 times brighter than a second-magnitude star, and a first-magnitude star is 6.25 (2.5 x 2.5) times brighter than a third-magnitude star.

Astronomers still use the apparent magnitude scale developed by Hipparchus, although with telescopes we can see many stars dimmer than magnitude 6. In fact, the Hubble Telescope can see all the way down to magnitude 29. Actually, that's magnitude +29, because while just about every star seen in the sky has a positive magnitude number, some are so bright they sport negative magnitude values. For example, Sirius, the brightest star in the night sky, is magnitude -1.42. Planets like Mars, Jupiter, and Venus can be as bright as magnitude -2 to -4. If you really want to get bright, the full moon is magnitude -12.6 and the sun is magnitude -26.8.

Remember, though, that this scale measures *apparent* magnitude, not actual magnitude. The brightness of a star you see is dependent on two factors: its distance from Earth and its luminosity, or how much light the star is putting out. If you place two light bulbs 100 yards away from you, and one is a 25-watt refrigerator light and the other is a 150-watt yard light, you know which will appear the brightest: the yard light. However, if you leave that yard light 100 yards away and move the refrigerator light a foot from your nose, the refrigerator light will appear the brightest. Scientifically, this effect is due to what is called the inverse-square law of light. Basically, the law says that if you double the distance of a light source, it will appear four times

Brocchi's Cluster, also known as "the Coathanger," in the constellation Cygnus (Photograph © Thomas Matheson)

as dim. If you triple the distance, the light will appear nine times as dim.

Astronomers have also developed a way to express the luminosity of a star. It's called absolute magnitude and it is the "real" magnitude of a star computed as if viewed from a distance of 32.6 light years. You have to mathematically push out or pull in each star using the inverse-square law of light. This method allows you to factor out distance when comparing the light output of stars.

The best examples of distance's effect on apparent magnitude are the stars Vega and Deneb. Vega is only 26 light years away while Deneb is over 3,200 light years away. Both Vega and Deneb are bright, but from our perspective, Vega appears the brighter of the two. However, if you were to magically pull Deneb in next to Vega, Deneb would be just about as bright as the full moon, and visible in the daytime.

4

STARS ON THE MOVE

Stars and constellations never stand still in the sky. They have perpetually itchy celestial feet and endlessly parade in giant circles around Polaris, the North Star. At the same time, from night to night many stars shift to the west, changing which constellations we see from season to season. The nightly and seasonal shift of the stars is due to the motion of the earth. Every stellar motion we see is just a reflection of the earth's 24-hour rotation on its axis and its 365.25-day orbit around the sun.

This extended-exposure picture captures the trails of stars as they move counter-clockwise on the celestial sphere around the North Star. (Photograph © Thomas Matheson)

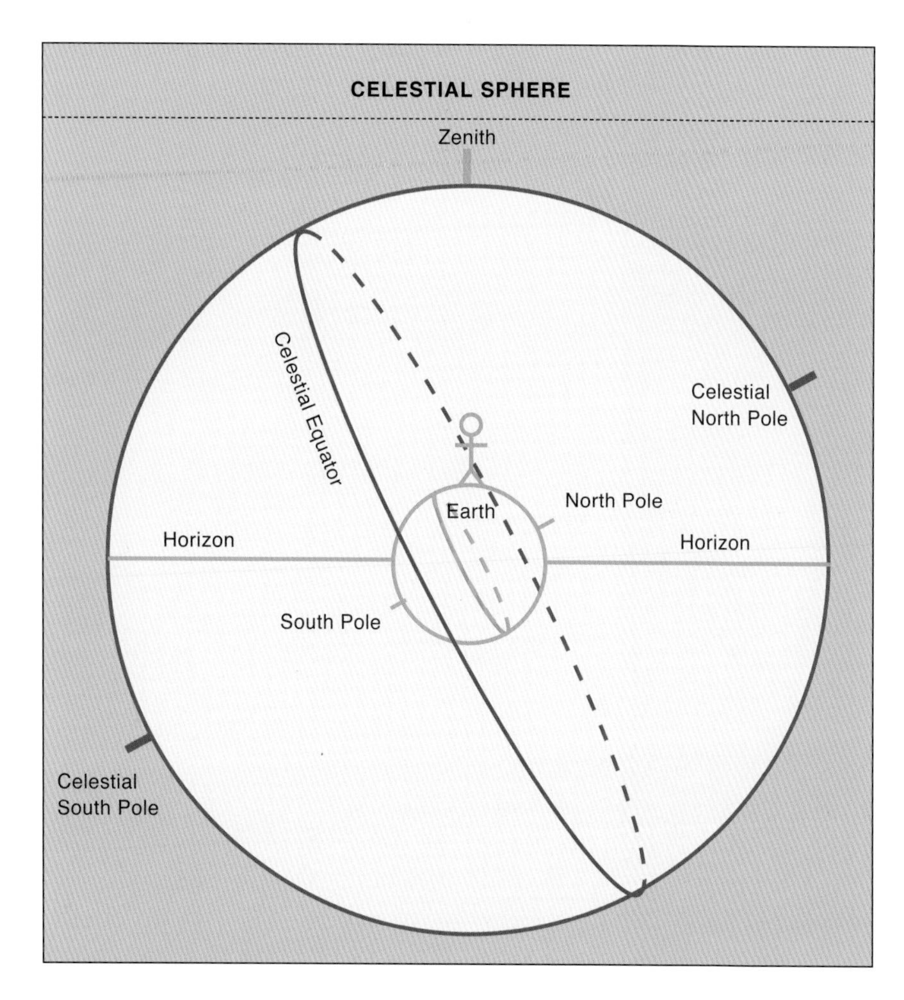

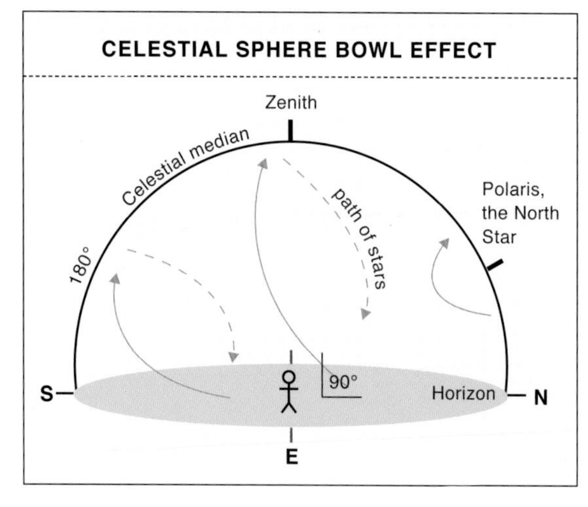

THE DAILY (AND NIGHTLY) MOVEMENT OF THE STARS

When you are out at night, it seems like a big bowl of stars is turned upside down over you. If you extended the bowl all the way around the earth, our planet would be surrounded by a giant bubble of stars, all fixed at the same distance from Earth. This is what the ancients called the celestial sphere and, although we know today that this idea of the universe is purely imaginary, it helps to explain the daily and nightly motion of the stars.

All around us is a 360-degree circle called the horizon, separating in all directions the land from this bowl of stars. If you look straight overhead, 90 degrees from the horizon in any direction, you're looking at what's called your local zenith. Your local celestial meridian is an imaginary line or half circle that runs 180 degrees from the northern horizon to the zenith, and then on to the southern horizon. It divides the sky in half, the western half and the eastern half.

If you sit back and face north—and you have the kind of weather that permits you to stay outside for a couple hours—you'll see the stars gradually turn counter clockwise around a point about one-third of the way up in the northern sky. You may notice a medium bright star staying stationary at this point. You're seeing Polaris, the North Star.

To confirm that it is Polaris, use the Big Dipper. Look for the two brightest stars, Dubhe and Merak, which make up the front lip of the pot section, opposite from the handle. Draw a line from Merak to Dubhe, then, holding your fist at arm's length, extend the line three more fist widths. This should guide your eye to the North Star.

Measuring arc distance in the sky is easy when you use your fist, fingers, and the moon. One fist-width at arm's length is equal to about 10 degrees of arc on the celestial sphere, so Polaris is about 30 degrees from Dubhe. One finger at arm's length is 1 degree. One full moon-width equals about half of 1 degree. As big as a full moon appears in the sky, you could actually line up 360 full moons along the celestial meridian.

Contrary to popular belief, the North Star is not the brightest star in the sky, though it is an important star. It's what I call the lynchpin of the sky, because every twenty-four hours all the stars appear to revolve around it. That's because the Polaris shines almost directly above the earth's North Pole, hence the nickname North Star. In actuality, Polaris is about 2 degrees away from the celestial north pole—a projection on the celestial sphere of the earth's North Pole—but that's close enough for stargazing.

If you were standing at the North Pole, hopefully with a warm coat and for heaven's sake a hat, you would observe all of the stars revolving around Polaris every twenty-four hours as the earth makes one full rotation on its axis. The North Pole would not only be a cold place to stargaze but also a boring place, because you would see the same stars every single night of the year, all whirling around the North Star at the zenith.

We live about two-thirds of the way from the North Pole to the equator, so Polaris in our sky is found about a third of the way between the northern horizon and the overhead zenith. The closer on the celestial sphere a star is to the North Star, the smaller its circle around it. The farther away a star is, the larger its circle around the North Star. Stars about 30 degrees from Polaris never fall below the horizon as they circle. At this latitude these stars are called circumpolar stars. The only true circumpolar constellation in the southern United States is Ursa Minor, the Little Bear. But we do see parts of Ursa Major and Cassiopeia the Queen every night of the year, barring those dastardly clouds.

Stars located more than 30 degrees from Polaris on the celestial sphere are called diurnal stars. Because they make larger circles around the North Star, they spend part of their circuit below the horizon. These stars rise in the east and set in the west, just like our sun and moon. Stars that are 30 to 90 degrees away from Polaris spend more time above the horizon than below it. They rise in the northeastern sky, reach their halfway point along the celestial meridian, and set in the northwest. Stars more than 90 degrees away from Polaris spend most of their time below our horizon, rising in the southeast and setting in the southwest.

Again, the daily motion in the sky is just a reflection of the earth's rotation and of where you are located on the earth. If you look at the full-scale model of the celestial sphere with the entire earth surrounded by a sphere of stars, you can tell that from our mid-northern latitude some stars visible to Earth never get above our horizon. Another helpful line of reference on the celestial sphere is the celestial equator, which is a projection of the earth's equator. It divides the north celestial hemisphere of stars from the south celestial hemisphere. Around here, we can see all the stars north of the celestial equator at one time or another, but a little more than half the stars south of the celestial equator. This is because of the curvature of the earth. The ground literally prevents us from seeing certain stars in the south celestial sphere.

If you travel to the equator, things change. You see more stars south of the celestial equator, and Polaris isn't as high in the northern heavens. The general rule of thumb is that the altitude of the North Star in your sky is equal to your latitude. If you get a deal on a flight to Hawaii, where the latitude is 21 degrees, you see Polaris at only 21 degrees above the horizon. At that latitude, the Big Dipper rises and sets like most stars. At the same time, you also see constellations in the southern sky that you can never see here, such as the Southern Cross.

MEASURING ANGULAR DISTANCE IN THE SKY

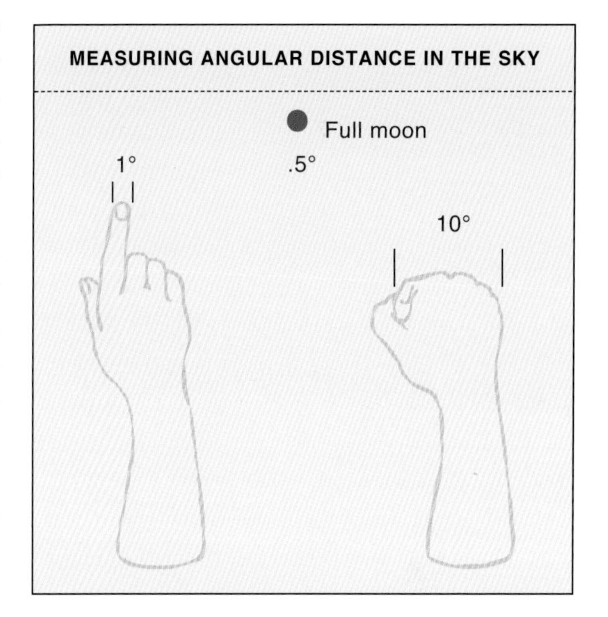

BIG DIPPER POINTER STARS

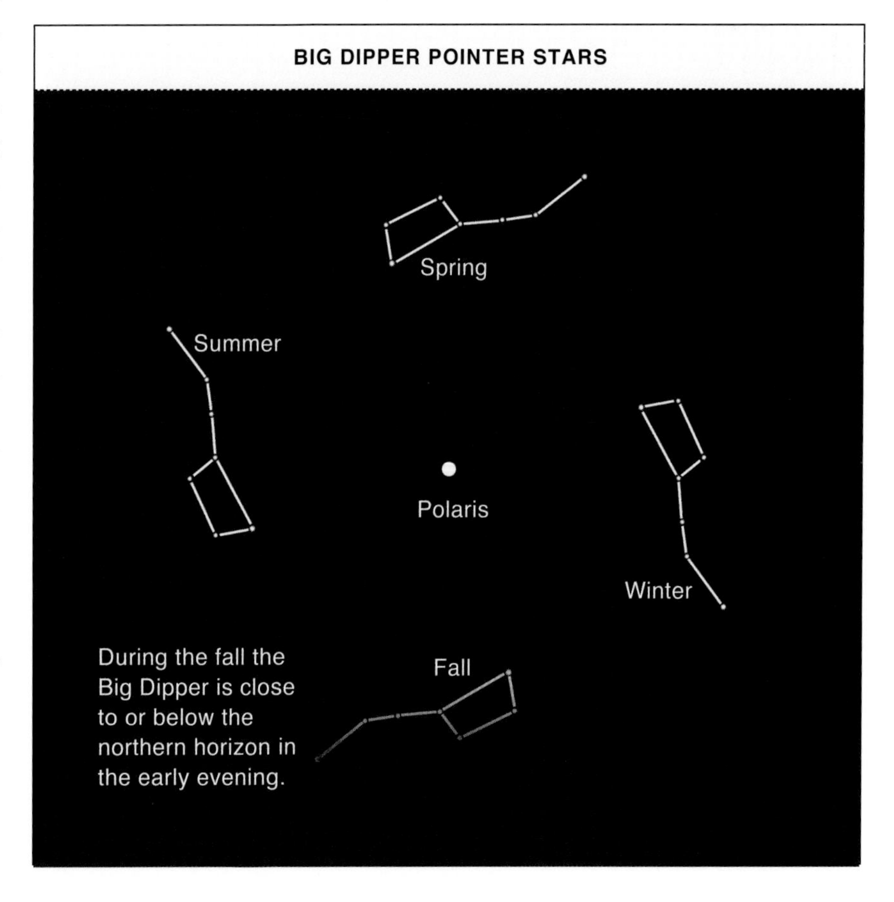

During the fall the Big Dipper is close to or below the northern horizon in the early evening.

If you traveled to Ecuador, which is located at the equator, the North Star would be on the northern horizon and there would not be any circumpolar stars. The payoff, though, would be that you would see all of the stars available to Earth. All eighty-eight constellations would pass over your head at one time of the year or another.

SEASONAL MOTION OF THE STARS

As the earth orbits the sun every 365 days, we see different constellations in different seasons. That's because Earth faces different directions in space as it circles the sun. For example, around eight o'clock at night in October we see the constellations Pegasus and Cygnus because our side of the earth faces the same direction of space as these constellations do. If we stargaze at eight o'clock at night in February, we see a different array of stars, including the constellations Orion and Canis Major, because we're looking into a different direction in space.

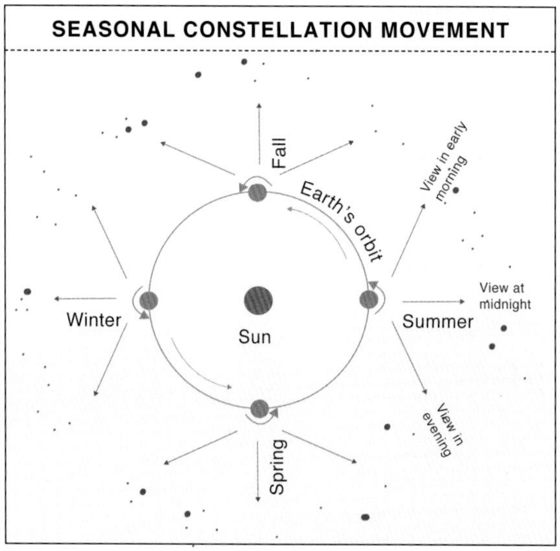

SEASONAL CONSTELLATION MOVEMENT

If you were to stargaze at the same time every night, after a few weeks you would notice that the constellations drift to the west each night as they circle Polaris. Over time, you would see new constellations coming up in the east as constellations in the west disappear below the horizon. Constellations rise and fall throughout the course of each night because of the earth's rotation, but this night-to-night westward drift is due to the earth's revolution around the sun. In addition to circling Polaris every twenty-four hours, all stars move an additional degree westward because of the earth's orbit around the sun. Over the course of ten nights, the stars drift 10 degrees, or one fist-width, west. Because of this drift, each star also rises and sets four minutes earlier each night.

The only exceptions are the circumpolar constellations, since they're always above the horizon. They shift their positions from night to night like all the other stars, shifting 1 degree counterclockwise around the North Star. Remember the adage "the early bird gets the worm"? Well, the early bird also gets a jump on the stars. The constellations visible in the morning sky are the same ones seen in the evening about four months later. If you roust yourself out of bed in early October at four o'clock in the morning, you'll see the same constellations visible at eight o'clock in the evening in early February. You're seeing the wonderful winter constellations in autumn. In a way, you're a season ahead of yourself!

THE ECLIPTIC

Along with the daily and seasonal motions of the stars on the celestial sphere, there's also the slower motion of the sun, moon, and planets to consider. The sun migrates on the celestial sphere on a specific path called the ecliptic. Remember, the celestial sphere is an imaginary creation and the ecliptic is simply the plane on the celestial sphere along which the sun appears to move. The sun makes one complete circuit along the ecliptic in one year, reflecting the time it takes the earth to orbit our sun.

The ecliptic is inclined to the celestial equator by 23.5 degrees because the earth's axis is tilted by that same 23.5 degrees as its whirls around the sun every year. That's why we have four seasons.

Around June 21, the average date of the summer solstice, the sun shines more over the earth's northern hemisphere than the southern hemisphere. Specifically, it shines directly over the tropic of Cancer, 23.5 degrees north of the equator. At noon on the first day of summer, the sun in our sky is at its greatest distance north of the celestial equator. It's the highest the sun ever gets over our area. On that day, we get more than fourteen hours of daylight, with the sun rising in the northeast, setting in the northwest, and taking a long arc across the sky.

Around December 21, the average date of the winter solstice, the sun shines more over the southern hemisphere than the northern hemisphere. On that day it shines directly over the tropic of Capricorn, 23.5 degrees south of the equator. In our sky at noon on the first day of winter, the sun is at its greatest distance south of the celestial equator. In other words, the sun's altitude is at its lowest noontime altitude of the year. On this day, we get only about ten hours of daylight, with the sun rising in the southeast, setting in the southwest, and taking a short arc across the sky.

During the vernal equinox in spring, around March 21, and the autumnal equinox in fall, around September 21, both hemispheres receive about the same amount of sunshine. On those days, the sun shines directly over the earth's equator, and is located directly on the celestial equator. The sun rises directly in the east and sets directly in the west. On the first days of spring and fall, the days and nights are nearly equal, about twelve hours each.

THE ZODIAC

As the sun makes its annual circuit along the ecliptic, it passes in front of twelve constellations, called the zodiac constellations. From month to month the sun passes eastward from one zodiac constellation to another. I'll bet you've heard names like Taurus, Scorpius, and Gemini before. The names of these constellations are the same as the twelve zodiac signs you read about in the horoscope column of your newspaper. Horoscopes are the centerpiece of astrology, which, despite a similar name, has nothing to do with astronomy. One is a science, and the other? Well, I have my opinion. How can the movement of the sun in front of a pattern of stars that are all at different distances and independent of each other have anything to do with whether or not your love life is going to improve, or if you're going to get that raise at work? It's a mystery to me! Of course, in my day job, I'm a meteorologist, so who am I to talk?

Astronomically, the zodiac is a 16-degree-wide band of the celestial sphere, with 8 degrees on either side of the ecliptic. It's where you find the zodiac constellations, the moon, and the planets. All the planets in our solar system, except Pluto, orbit around the sun in nearly the same plane. If you could put the entire solar system in a giant clothes dryer and shrink it to the size of your kitchen table, the planets and the sun would lay nearly flat on the place where you have breakfast every morning.

The orbit of the principal planets can vary by over 7 degrees compared with the earth's orbit. On the celestial sphere, this means the planets can be more than 7 degrees on either side of the ecliptic, so the 16-degree-wide zodiac band easily accommodates this variance. Pluto's orbit is inclined to the earth's orbit by more than 17 degrees, placing this distant renegade planet mostly outside the zodiac band of constellations.

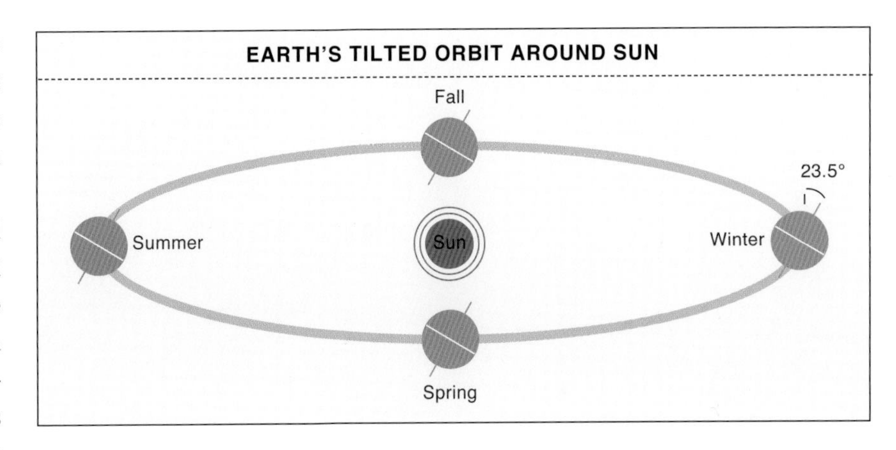

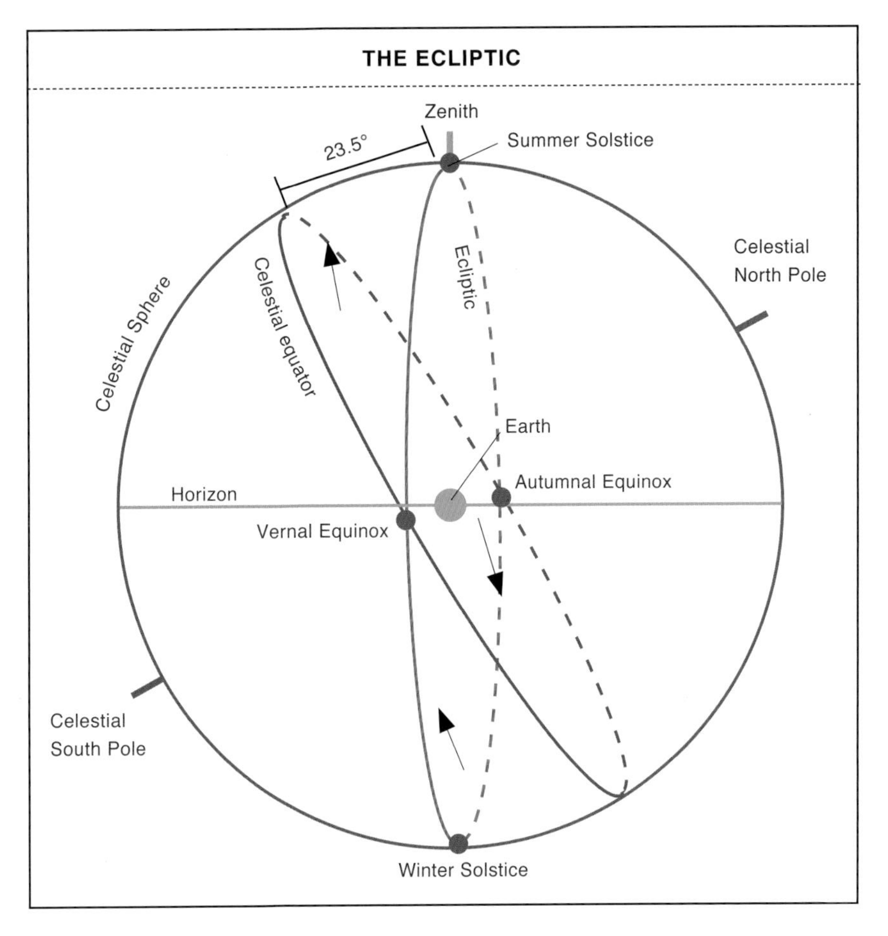

5

USING STAR MAPS

It's time to get outside and make the stars your old friends. Now that you know more about stars and constellations and how they move across the sky, take this book under the great star dome and have some fun. In the appendix, you'll find user-friendly star maps for each month of the year. All kinds of celestial goodies await you!

Eagle Nebula, M16, in the constellation Sagittarius (Photograph © Robert Gendler)

You might be wondering how you are supposed to see these maps out there in the dark. A regular white flashlight is no good; you'd see the maps all right, but you wouldn't see the stars because you'd be night blinded. You know how it is when you're out at night. You start seeing better after ten to fifteen minutes in the dark. Your eyes become nocturnal. Your pupils widen to gather more light in the darkness. If you switched on a white flashlight to look at the star maps, your pupils would quickly contract and your night vision would become all loused up. A word of caution here: if you go to a public viewing party put on by an astronomy club, never find your way around with a white flashlight. They're great people, but they are hardliners when it comes to preserving their night vision.

The solution is red light. A moderate red filtered flashlight gives you the light you need and your night vision will not be affected. You can buy a special red flashlight at a store that sells telescopes, or you can save money and make one yourself. Just take an old flashlight and paint the lens red with model airplane paint, or cut out a piece of red construction paper and fit it into the lens. Recently I've started using one of those little flashlights that attaches to headgear. It's great because it frees up my hands for star charts and binoculars. You can find these "headlights" at just about anyplace that sells camping gear.

Checklist For Successful Stargazing

- Star maps and constellation charts
- Red-filtered flashlight
- Small pair of binoculars (7 x 35)
- Reclining lawn chair
- Snacks and something to drink
- Blankets and sleeping bags
- Radio or CD player
- Friends and family!

A telescope is an added bonus but not essential for a great night of stargazing. Just be ready to have a good time and make the stars your old friends!

Once you're equipped with the proper lighting, where do you use these maps? Naturally, the darker the location, the better. If you live in the country, you're good to go. If you live in the city or suburbs, consider a road trip; head to a park, campsite, or backcountry road. If you know a farmer out in the boonies, that's even better. My advice is to know where you're going ahead of time and make arrangements if necessary. Driving around in the dark looking for a place can be frustrating and time consuming.

If you don't have time to take your stargazing on the road, you can still see a lot of good celestial stuff; you'll just have to work at it harder. Mid- to outer-ring suburbs can have very friendly skies. Even in metro areas, you'll see a lot more than you think. Just stay far away from shopping malls and street lights, and learn to use averted vision.

Be sure to bring along a comfy lawn chair, preferably a chaise lounge that allows you to lean back and take in the stars. Even if it's winter, get those chairs out of storage. My wife likes to use neck pillows, the kind used when flying long distances. Have a thermos of your favorite beverage, or in the warmer months, a cooler. Finally, have some heavy blankets or sleeping bags on hand so you can stay out awhile. Stargazing is great year round, especially in winter. I think that's the best show of the year.

Averted or peripheral vision is a good tool for stargazing, especially if your skies are compromised by light pollution. Because of the structure of the eyeball, the edges of the eyes are more sensitive to incoming light than the middle. That's why you see quick motion so well out of the corner of your eye. When you're trying to see a faint constellation, turn your head slightly away and look at that area out of the corner of your eye. It works!

HOW TO USE THE MAPS

The maps are easy to understand and easy to use. You'll find a star map for every month of the year, set for the stars' positions in the middle of the month, but usable anytime during the given month. The maps represent early evening skies, generally about an hour and a half after sunset. These maps are designed for 30° north latitude, but they can be used between 25° and 40° latitude. If you live near the southernmost reaches of this range, the constellations in the northern sky will be a tiny bit lower than on these star maps, and the constellations in the southern sky a tiny bit higher. The farther north you live, it'll be just the opposite.

As an example of the monthly star maps, look at the map for October. Everything inside the circle appears in the sky that month. The circle represents the horizon. The middle of the map is your overhead, or zenith. Maybe you are wondering why the eastern horizon is on the left and the western horizon on the right. Every other map you've seen is just the opposite. Is this a misprint? Is the author of this book a little backward? Not in this case. East and west on the star maps are in their proper places as long as you hold the maps the way you should, that is, over your head. When you do that, I guarantee that east and west will point where they should.

Holding the map over your head and simply shifting your eyes from map to sky is much easier than looking down at the map and bobbing your head up to look at the stars. That's doing it the hard way and what I call "bobble head" stargazing!

The best technique for using the star maps is to take on one direction of the sky at a time. Start with the northern sky. Sit in your lawn chair facing north. Hold the map over your head so that the north point on the horizon circle faces north. Then look for everything that can be seen in the northern sky, like the Big and Little Dippers, Polaris, and the constellations Cassiopeia the Queen and Cepheus the King. When you've found all you can in the northern sky, turn your chair west. Resume lounging, holding the map over your head with the west point on the horizon circle facing west. Then find all you can in the western sky, like Boötes the Farmer and Hercules the Hero. Repeat this procedure for the southern and eastern skies until you've made a full circle.

With a little patience and persistence, I think you'll find this method works best for using the maps. What you don't want to do is take on the whole sky from one sitting position or you'll soon be seeing a chiropractor for neck problems. The monthly maps are great for beginners, as well as stargazers who have to put up with moderate light pollution, because it only shows the brighter stars and constellations. The dimmer and less significant constellations are filtered out to keep this map simple.

The maps depict the early evening sky for their respective months, but they may also be used other months at other times. For example, you can use the October sky map at four o'clock in the morning in June or midnight in August.

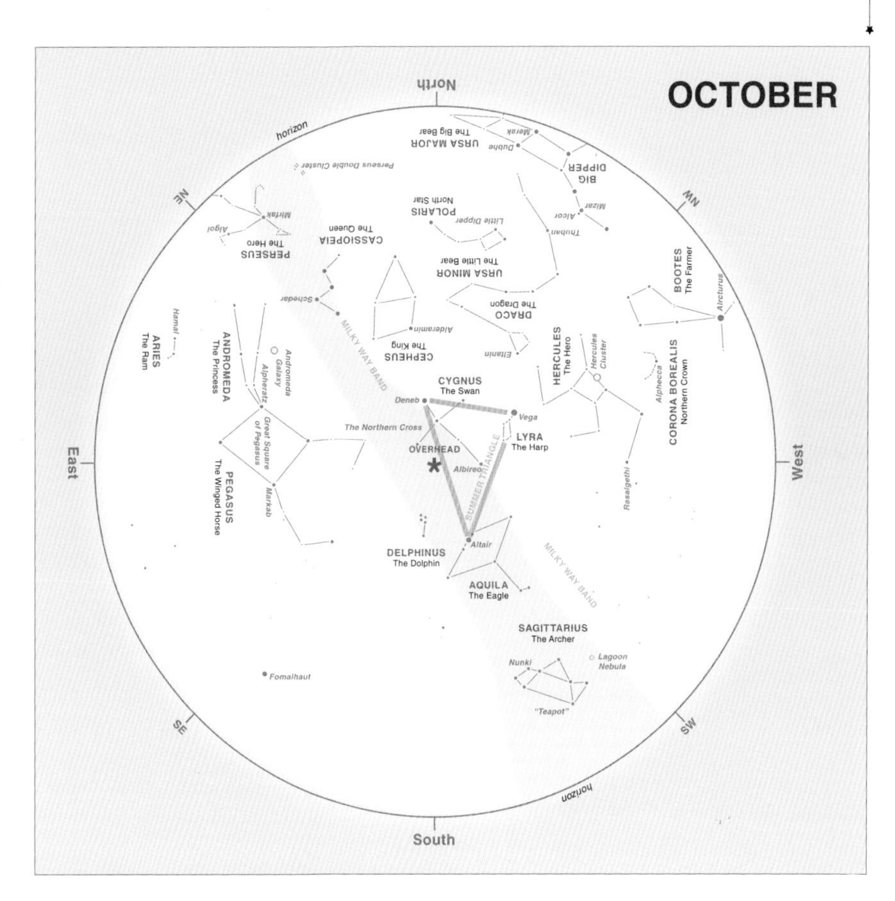

CELESTIAL GOODIES

Our night sky contains lots of hidden and not-so-hidden treasures. Among the constellations lurk double and triple stars, variable stars, star clusters, nebulae, and even other galaxies. Most of these require telescopes or binoculars, but some of these treasures can be seen with the naked eye. The brighter celestial goodies are labeled on the monthly star maps and the not-so-bright ones are plotted on the individual constellation charts in the next chapter.

On the star maps, you'll see objects denoted by "M" numbers like M13 or M42. These are "Messier objects," named after Charles Messier, a French astronomer who lived in the eighteenth century. He was one of the first "comet hunters." When comets first show up through a telescope, they're just small, fuzzy patches that slowly move among the background stars. Messier wanted to be the first to see Halley's comet on its return in that century. Part of the thrill of finding new comets is that they're named after their discoverers. But he kept finding fuzzy patches in the sky that didn't move. Too often he confused them with potential new comets.

Messier wasn't interested in these stationary patches, but he cataloged their positions so he wouldn't confuse them with comets. He came up with a list of over a hundred Messier objects that is still in use today, only now we care about what these Messier objects actually are. They are star clusters, nebulae, and other galaxies.

NGC OBJECTS

In 1888, J. L. E. Dreyer, an astronomer based in Northern Ireland, published a more complete list of nebulae and star clusters, titled "The New General Catalogue of Nebulae and Clusters of Stars." The name was later shortened to New General Catalogue or NGC. This catalog contains over 78,000 objects observed by several astronomers, including William Herschel, who discovered the planet Uranus.

NEBULAE

Nebula is a Greek word that means cloud and that's exactly what nebulae are, vast clouds of interstellar hydrogen gas and dust that are all over our galaxy and universe. There are several types of nebulae: dark, bright emission, reflection, planetary, and supernovae.

Dark Nebulae

Most nebulae are dark clouds that can't be seen directly. If no bright stars are nearby, nebulae are just big, dark masses that only show up when they partially block our view of objects beyond them. Lots of dark nebulae block our view of bright starry fields that would otherwise show up in our night sky. Dark nebulae also block our view of the center of our galaxy, in the direction of the constellation Sagittarius. Otherwise, astronomers believe that this star-congested area would be brighter than the full moon!

Bright Emission Nebulae

Bright emission nebulae glow like giant neon lights. Emission nebulae are gigantic celestial nurseries, and the sources of their light are the extremely hot, bright stars born out of them. If the newborn stars are massive enough, they produce sufficient intense ultraviolet radiation to ionize the atoms in the nebula. Electrons are constantly bouncing away from and back to the hydrogen atoms, causing the lovely glow we see. In backyard telescopes, most emission nebulae give off a greenish glow, but photographs from larger scopes and the Hubble Telescope also show reds and yellows.

The best emission nebula in the night sky is the wonderful Orion Nebula, M42, in the constellation Orion the Hunter. It's easily seen with the naked eye in the sword of the hunter, over 30 light years in diameter and around 1,600 light years away. With even a small telescope, you can see four young stars, arranged in a tiny trapezoid in the middle of the Orion Nebula. Some of the infant stars may be less than a million years old.

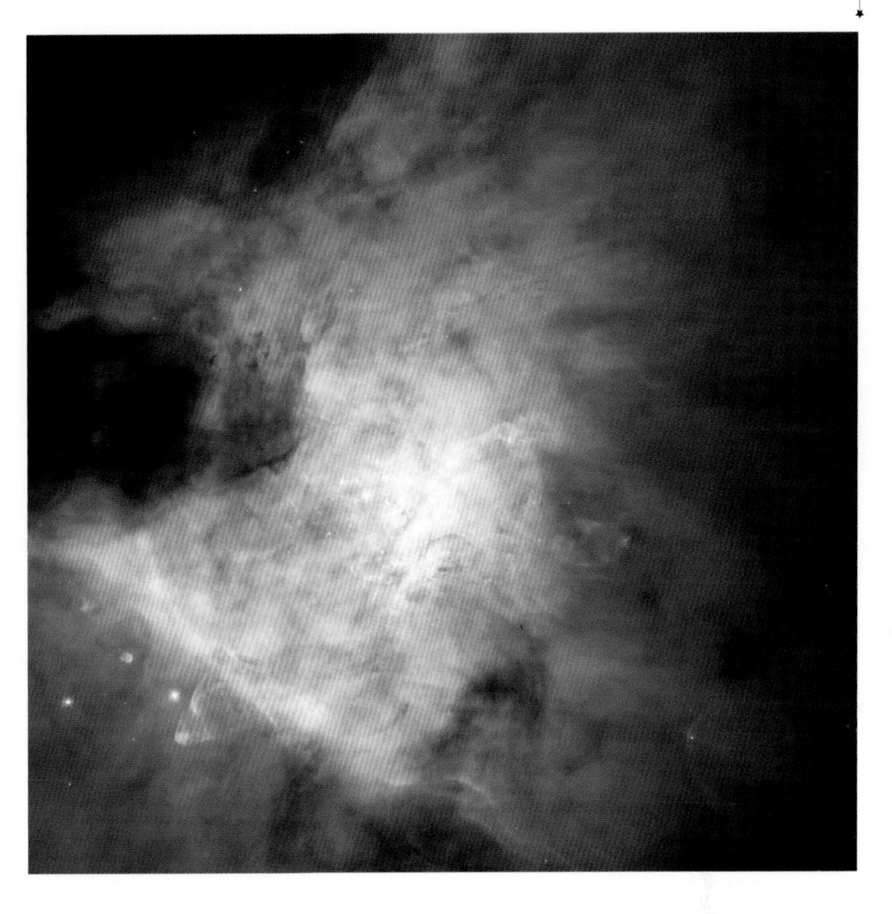

ABOVE: The Orion Nebula, M42, is a bright emission nebula in the constellation Orion. (Courtesy of C. R. O'Dell, Vanderbilt University, and NASA)

FACING PAGE: The Horsehead Nebula is a dark nebula in the constellation Orion. (Photograph © Robert Gendler)

Reflection Nebulae

These aren't as flashy as emission nebulae but still put on a good show. Reflection nebulae are mainly clouds of hydrogen gas and dust that surround newly formed stars. Unlike emission nebulae, in which the gas is being excited by the newborn stars, the gas and dust in reflection nebulae merely reflect starlight from young stars. They're generally bluish-white and not as easy to see as emission nebulae, but with a telescope you'll have some success. The best example of a reflection nebula is the gas that surrounds the giant Pleiades Star Cluster, M45, seen in the evening sky from late fall to early spring.

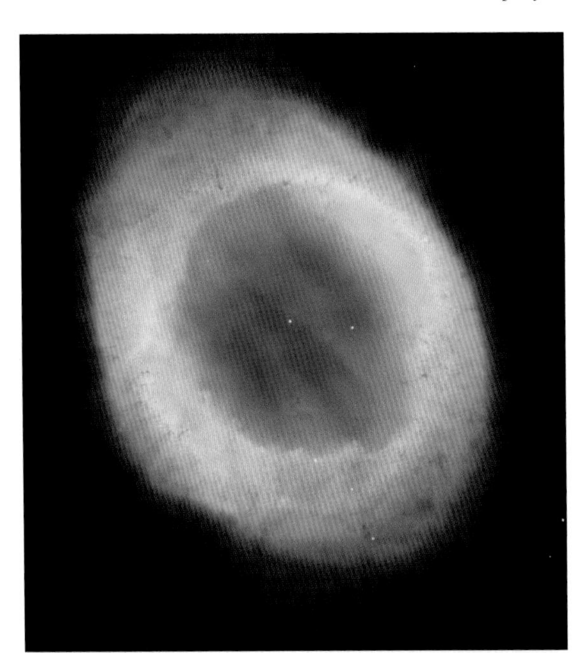

The Ring Nebula, M57, is a planetary nebula in the constellation Lyra. (Courtesy of NASA and the Hubble Heritage Team (STScI/AURA))

Planetary Nebulae

This type of nebulae have nothing to do with dark, emission, or reflection nebulae. Planetary nebulae result from dying stars burping off shells of gas as they evolve from red giants to white dwarfs. They are called planetary nebulae because they resemble a planet with rings, just like Saturn. You need a decent telescope to see them. One of the brightest is the Ring Nebula, M57, over 2,300 light years away, located in the constellation Lyra the Lyre. Through the eyepiece, it looks like a ghostly bluish-white smoke ring.

Supernovae Explosion Nebulae

Supernovae nebulae are similar to planetary nebulae in that they're made from the gas of dying stars. In the case of supernova clouds, however, you are looking at the remnants of a supernovae explosion. The magnitude of this explosion makes the biggest hydrogen bomb explosion look like a firecracker. For a brief time, the exploding star may be as bright as an entire galaxy. The best example we have of a supernova remnant is the Crab Nebula, M1, in the constellation Taurus the Bull.

STAR CLUSTERS

All around a dark starlit sky, even with a casual glance, you see clumps and clusters of stars. With binoculars and telescopes, you see many, many more. Most of these star clusters are open clusters, giant nuclear families of young stars. There are also globular clusters, which are made of some of the oldest known stars in the universe.

Open Clusters

These are the infants of the universe. Just about every star in the heavens was once a member of a large open cluster of young stars, born out of a large nebula. These clusters stay together because of gravitational attraction between the sibling stars. After several orbits around the galaxy, the Milky Way in our case, gravitational attraction from passing stars breaks up the clusters. Many stars go off on their own, sometimes as single stars like our sun, or in double, triple, quadruple, or higher-numbered star systems.

There are many good examples of open star clusters in our night sky. I already mentioned the Pleiades, made up of over 100 stars over 400 light years away. The stars in the Pleiades are estimated to be only 80 to 100 million years old. Compared with other stars, they're not even out of diapers!

Other great clusters include the Beehive Cluster, M44, in the spring sky near the constellation Leo the Lion; the Wild Duck Cluster, M11, just off the tail of the constellation Aquila the Eagle; and the wonderful Perseus Double Cluster in the constellation Perseus the Hero.

Globular Star Clusters

While open star clusters are nurseries for baby stars, globular clusters are the crowded nursing homes for old stars. Globular clusters are extremely tight spherical clusters from 50 light years to over 300 light years in diameter, crammed with anywhere from hundreds to up to 2 million stars! A cluster is held together by the mutual gravity of all its stars. Astronomers think globular clusters may have formed when the universe was very young, around the time of the formation of the Milky Way or possibly even before. That means they might be over 13 billion years old. For reasons we don't totally understand, these ancient clusters form huge halos around galaxies, including the Milky Way. They can be wonderful through a telescope. At first glance, they look like little fuzz balls, but with some focusing, their true beauty splashes out at you.

The best globular cluster in the night sky is the great Hercules Cluster, M13. It is located near the center of the faint constellation Hercules the Hero and, even with a small telescope, I know you'll be pleased. M13 is 25,000 light years away and about 140 light years in diameter. It's estimated that over half a million stars are shoehorned into this ball. With the telescope, you can see some of the individual stars at the edges, but in the middle so many stars are so close together that it's one big mass of light with a bluish tinge.

GALAXIES

In the late 1920s and '30s, Edwin Hubble made the startling discovery that some of the fuzzy spots thought to be nebulae were actually separate galaxies, millions of light years beyond our Milky Way. So many Messier and NGC objects are now known to be galaxies. For example, prior to Hubble's discovery, M31 was known as the Andromeda Nebula. Now it's the Andromeda Galaxy, 2.5 million light years away and the next-door neighbor to the Milky Way. It's also the most distant object the human eye can see. If you're stargazing in the dark countryside, you can see it with the naked eye, or at least with binoculars or a small telescope.

Astronomers think there could be over 100 billion other galaxies besides our own in the known universe, some possibly over 13 billion light years away! The Hubble Telescope has photographed many of the distant ones, but with a backyard telescope we can see many of our close galactic neighbors.

A reflection nebula surrounds the open star cluster Pleiades, M45, in the constellation Taurus. (Photograph © Rick Krejci)

<u>FACING PAGE, TOP:</u> The Crab Nebula, M1, is a supernova nebula in the constellation Taurus. (Photograph © Robert Gendler)

<u>FACING PAGE, BOTTOM:</u> The Hercules Cluster, M13, is a globular cluster in the constellation Hercules. (Photograph © Robert Gendler)

<u>TOP:</u> The Wild Duck Cluster, M11, is an open cluster in the constellation Aquila. (Photograph © Russell Croman)

<u>BOTTOM:</u> The M33 is a small spiral galaxy in the constellation Pegasus. (Photograph © Robert Gendler)

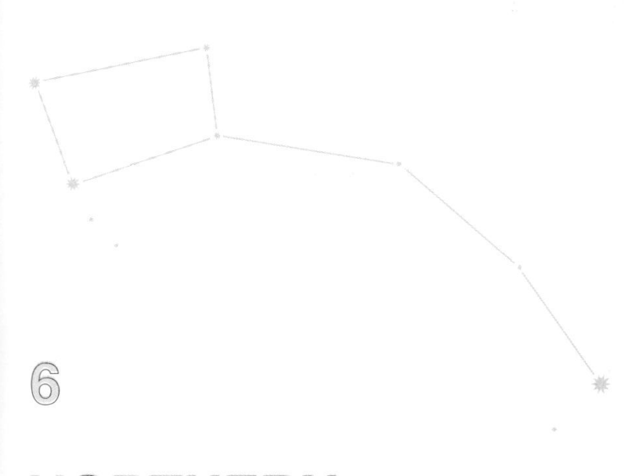

6

NORTHERN CONSTELLATIONS

Our most faithful friends in the night sky are the northern constellations. The brightest ones–Ursa Major, Ursa Minor, Cassiopeia, and Cepheus–are at least partially visible above the horizon every clear night of the year, faithfully circling the north celestial pole. Some of the stars in these constellations dip slightly below the northern horizon for a short time as they circle the North Pole.

Most constellation lore comes from ancient Greek and Roman mythology. Oftentimes the stories interchange Greek and Roman gods and, to really make things more confusing, any one constellation can be associated with up to a dozen stories. I'm not going to bore you with all of them, but I have picked out the ones I think are the best. I have to warn you, though; I've been known to change these stories here and there, all in good fun and hopefully to make them more memorable.

Bode's and Cigar Galaxies, M81 and M82 (Photograph © Robert Gendler)

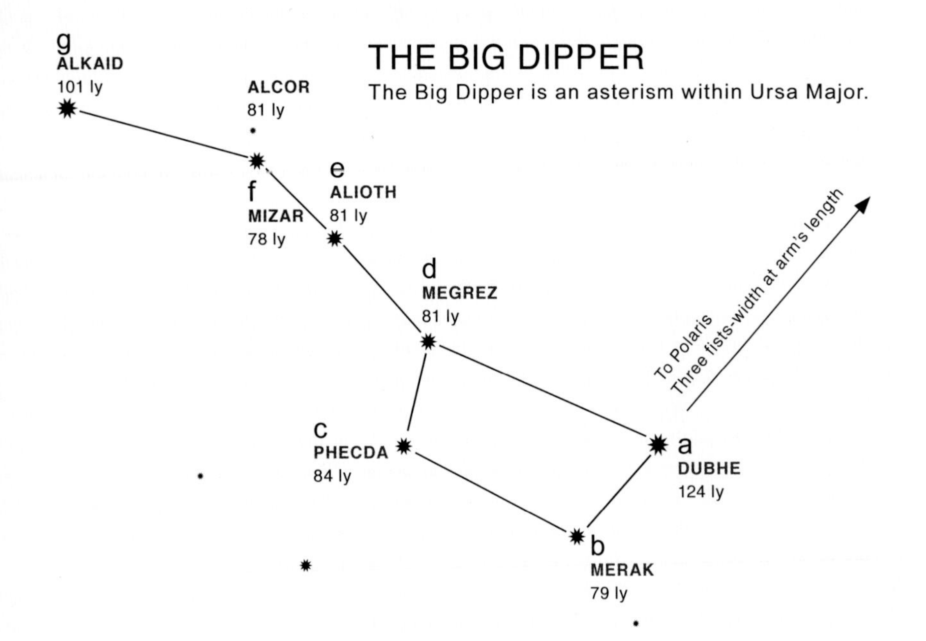

THE BIG DIPPER

The Big Dipper is an asterism within Ursa Major.

Key characters in Greek mythology include Zeus, the king of the gods, and his wife Hera, the queen of the gods. Their marriage is no model for young couples. You could say they have the classic dysfunctional relationship. Other characters also make regular appearances in these soap operas of the night sky. Apollo, the god of the sun, leads a chariot holding the sun and pulled by four flying horses across the sky. His twin sister is Artemis, the goddess of the moon. There's also Athena, the goddess of wisdom; Poseidon, the god of the sea; Ares, the god of war; and many more.

Hopefully you will return to these constellation chapters over and over again, sometimes reading them outside with a red light. The constellation charts include what I call celestial goodies, most of which are visible only with a telescope. These goodies are nebulae, clusters, and galaxies—the Messier and NGC objects. Have fun and be patient. It will take you many nights to get to know the constellations and their telescopic treasures. Stargazing is a long-term pleasure!

THE BIG DIPPER

The Big Dipper is, without a doubt, the most recognizable star pattern in the heavens. Even people who couldn't care less about astronomy know the Big Dipper. If I pointed you north and directed your eyes to the sky, you would find the Big Dipper without too much trouble, unless you're competing with mall parking-lot lights.

Most star and constellation lore was brought to America from Europe, but the Big Dipper is one of the few exceptions. Across the Atlantic, what we call the Big Dipper has been called many other names. In England, this grouping of stars is seen as the plough. In Germany, it is Charles's wagon. In Ireland, it takes on a biblical theme, as King David's chariot.

Before colonists descended on America, some Native American tribes saw these stars as a bear on the run from a family. The four stars we see as the dipper's pot section represent the bear and the handle the members of the family. The closest star to the bear is the father with a bow and arrow, the next star is the mom with the cooking gear, and tagging behind is the one of the kids.

The invention of the name Big Dipper, by most accounts, involves a sad but significant era in our history. African-American slaves, during and prior to the Civil War, saw this pattern of stars as a drinking gourd. Slaves drank water from hollowed out gourds, similar to the ones we see as Thanksgiving decorations. They paid special attention to the giant drinking gourd in the sky because it was always in the northern heavens, in the direction of the Union states, where freedom awaited. After the war when slavery was abolished, the drinking gourd in the sky evolved into what we now call the Big Dipper.

Constellations are mainly accidental scatterings of stars that happen to appear in the same general direction of space. Physically the stars have nothing to do with each other. One exception is the Big Dipper. Five of its seven stars are believed to have formed in the same nebula, beginning their stellar life about 200 million years ago as a small cluster that has been breaking apart ever since. More than thirty other stars in the sky used to be part of this same cluster. Only the Big Dipper's two outermost stars, Dubhe and Alkaid, did not come from this cluster. The rest of the stars are 80 light years away, give or take some.

There's a wonderful natural eye test in the Big Dipper. Look for the double star Mizar and Alcor in the handle. Mizar is a bright star, while Alcor is much dimmer. In ancient times, this double star was given the title Horse and Rider, and some Arabic literature refers to it as the Riddle. If you can see Alcor, your long-range vision is great; if you can't, it's time for a new prescription.

Mizar and Alcor are what is known as an optical double star. Whereas some doubles, called binary stars, orbit about a common center, optical doubles simply line up together in our sightline. Although these stars look to be right on top of each other, Mizar is 78 light years away and Alcor is 81 light years distant. If you look at Mizar through a telescope, you see that it's actually a binary star in which the two stars follow a slow orbit around each other. In the late 1800s, spectroscopic analysis showed that each of the double stars in Mizar has its own companion, making Mizar a double-double binary star.

There's a lot to say about the Big Dipper, but there is one small detail that shouldn't be overlooked. It's not a constellation! You would think it would be but it's not. It's what astronomers call an asterism. Worse yet, the Big Dipper has been relegated to serve as the rear end and tail of that giant bear in the sky, Ursa Major.

URSA MAJOR, THE BIG BEAR

The constellation Big Bear takes up a lot of real estate in the northern heavens. The Big Dipper alone has some size, but it only makes up a third of Ursa Major. While just about everyone in the northern hemisphere has seen the Big Dipper, most people can't identify the entire bear. That's because the Big Dipper shines as the brightest part of the beast.

With dark-enough skies, you can visually hunt for the rest of the bear, especially in the spring when it rises in the northeastern sky. First, find the Big Dipper. Keep in mind that its handle is the bear's tail and its pot is the bear's hindquarters. Beyond the front lip of the pot, look for three faint stars that make a skinny triangle. They form the bear's neck and head, with the star Muscida at the nose. After that, look below the bear's head for two stars that are close together, almost what you would call a double

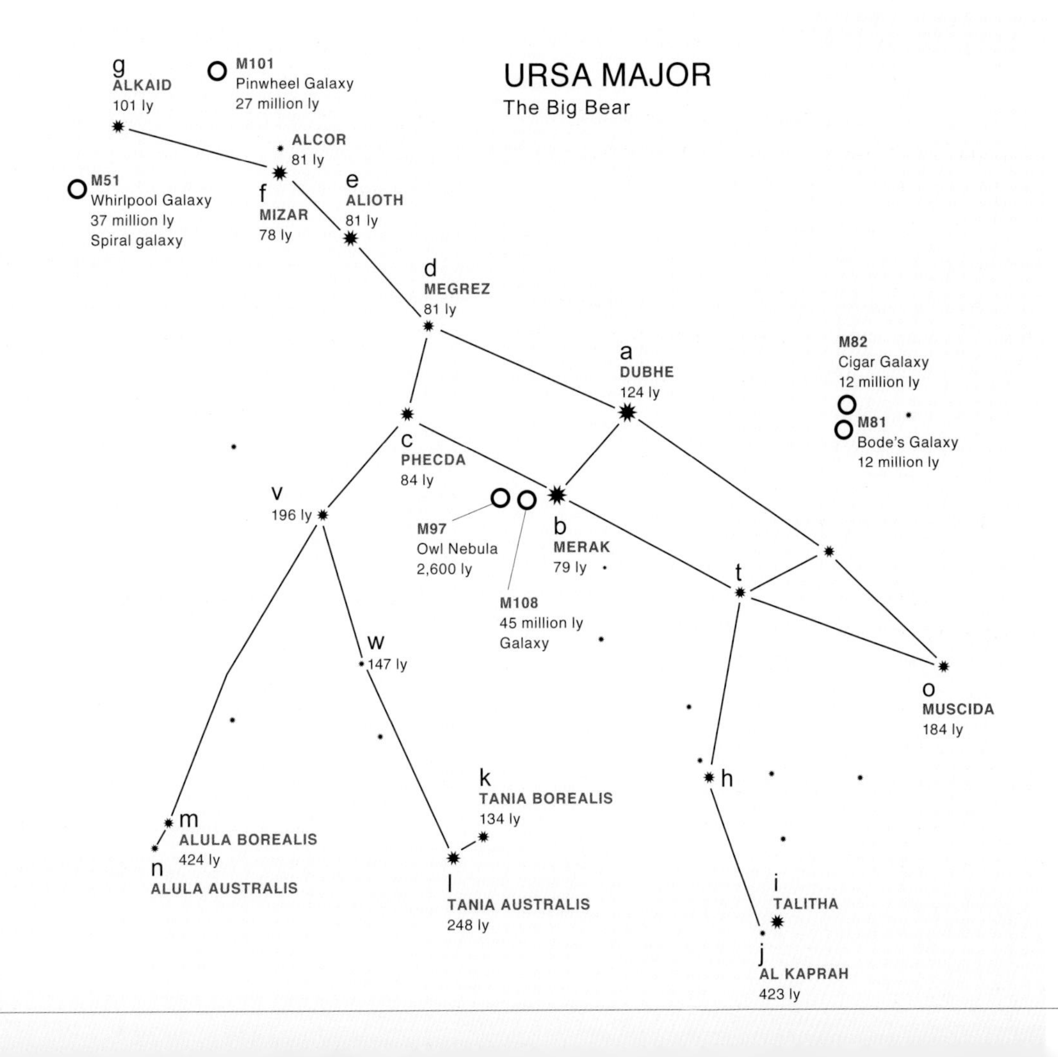

star. Those stars, Talitha and Al Kaprah, form one of the bear's front paws. The bear's front leg curves back to Upsilon Ursae Majoris, part of the bear's triangular head. Be sure to trace your imaginary line along the star between the front paw and the head that marks the front knee of Ursa Major.

Pinwheel Galaxy, M101 (Photograph © Russell Croman)

The best way to find the bear's back leg is to find the back paw. Just like the front paw, it is comprised of two stars, Tania Borealis and Tania Australis, that shine right next to each other. They aren't as close together as the front-paw stars, but they stand out enough. From the back paw, follow a crooked line of two more stars until you find your way to the star Phecda on the rear end of the bear.

The Big Bear is one of the grandest constellations of the sky. Locating this constellation will give you the visual skills needed to hunt down even fainter constellations.

Celestial Goodies

Mizar and Alcor: Together, these stars make up an optical double star. They are separated in the sky by less than a fifth of a degree. Mizar is actually a double-double binary star.
Magnitude: 2.2 and 4.0

M81 and M82: The most famous pair of galaxies in the sky, Bode's Galaxy, M81, and the Cigar Galaxy, M82, can be picked up even by a small telescope. We see the spiral-shaped Bode's Galaxy face-on, but the Cigar Galaxy edge-on, giving it the look of a fat cigar. The galaxies are separated by only 150,000 light years.
Magnitude: 7.8 and 8.4

M101: This large spiral galaxy, called the Pinwheel Galaxy, is 170,000 light years in diameter, almost twice as big as our Milky Way. We see it face-on from Earth, although low surface brightness makes it tough to see, even with a moderate to large telescope. It's best seen during spring and summer evenings, when it's high in the sky.
Magnitude: 8.2

M97: If you have clear skies, this planetary nebula, dubbed the Owl Nebula, really does look like the face of an owl. It is faint, however, and you'll need a moderate to large telescope to see it. As with all the celestial goodies in Ursa Major, the Owl Nebula is best seen during spring and summer evenings.
Magnitude: 12.0

M108: This faint galaxy, seen edge-on from Earth, is right next door to the Owl Nebula.
Magnitude: 8.6

M51: The Whirlpool Galaxy is located just below the star Alkaid. We look over the top of the galaxy, which may contain 100 billion stars in an area 50,000 light years across. It is one of the best galaxies to view. With a small to moderate telescope, you can even see its spiral arms.
Magnitude: 8.0

URSA MINOR, THE LITTLE BEAR

The Little Dipper and the constellation Ursa Minor are one and the same. Latin for "little bear," Ursa Minor is much easier to see as a dipper than a bear, but a word of warning: the Little Dipper is much fainter than the Big Dipper. In fact, those who stargaze around heavy city lighting often barely see it, and some can't see it at all.

There is a good strategy for finding the Little Dipper. First, draw a line from Merak to Dubhe in the Big Dipper, then extend that line three fists-width to arrive at Polaris. The North Star marks the end of the Little Dipper's handle.

Once you're locked on Polaris, find the closest bright stars. They should jump out at you. Kochab (pronounced ko-cab) and Pherkad (pronounced per-kad) make up the outside lip of the pot. Your mission is to find the other two pot stars and the two faint stars that create the curved handle leading back to Polaris.

Once you see the Little Dipper, load up your imagination to see it as the Little Bear. The pot is its body and the handle is its long tail.

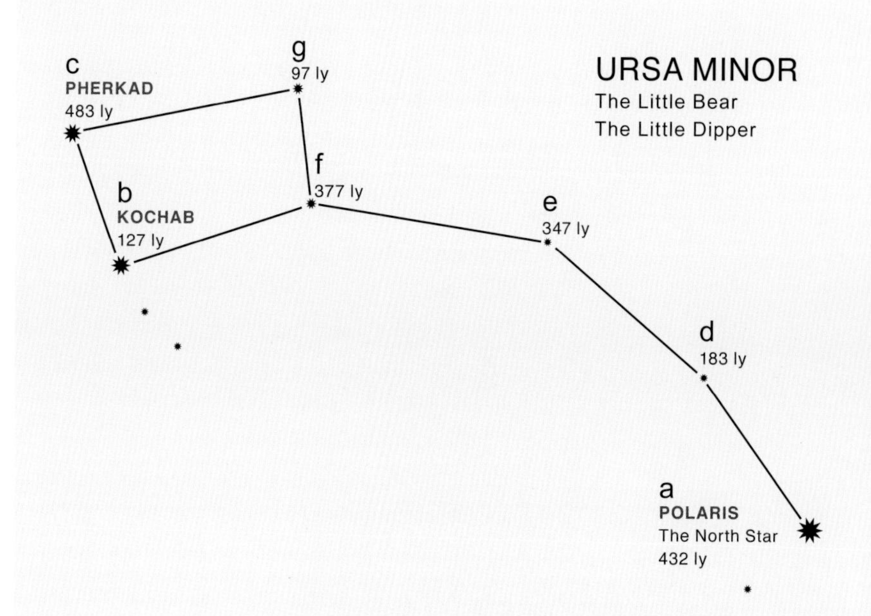

Celestial Goodies

Polaris: Located less than 1 degree from the north celestial pole, the North Star is the lynchpin of our sky.

THE "TAIL" OF THE BEARS

The story of how the bears got in the sky is one of my favorites. In Greek mythology, Callisto was a beautiful woman, who tragically had been left a widow and single mom several years earlier, when her husband was killed in war. Arcus, Callisto's wonderful eight-year-old son, helped his mom around the house and made life a little easier.

There were tough times, though, like the day Callisto's washing machine broke down and the only Laundromat within 50 miles was closed for remodeling. The young widow was forced to wash her clothes by hand in a nearby lake. As Callisto toiled over her laundry, Zeus, the king of the gods, happened to stroll by. He was legendary for being a ladies' man. Zeus gave Callisto his million-dollar smile and helped her wash the clothes. He then talked her into a cup of coffee at a local bistro. For the next few days, Zeus showered Callisto with attention and gifts, and they made plans to go hiking in the park the following Saturday. All this would have been fine except Zeus was already engaged to marry the goddess Hera.

One thing you don't do is anger Hera. When she heard about Zeus's love adventures, she was furious. After all, she already had the flowers ordered and the VFW hall reserved.

Hera traveled to the park early that Saturday morning and hid in the bushes. Sure enough, later that morning, along came Zeus and Callisto, arm in arm. As they approached, Hera jumped from the bushes, pointed her magic finger at Callisto, and turned her into a bear. Hera then dragged Zeus away by his ear.

Arcus was now without a mother as well as a father. He didn't know what had happened to his mother, just that she had disappeared. But life goes on. Arcus moved in with relatives and grew to become, of all things, a professional game hunter.

One day, as Arcus was trying out his new bow and arrow in the woods, a giant female bear sauntered his way. Yes, it was the same bear Hera conjured several years earlier, but Arcus didn't know that. He raised his bow to shoot his own mother. Talk about Greek tragedy! Fortunately, Zeus was once again out for a stroll and happened upon the scene, and he instantly recognized the bear from that ugly Saturday morning. He tried to convince Arcus that the bear was his mother, but Arcus didn't believe him and took aim at his mom. Zeus couldn't let this happen. Out of desperation, just before the arrow flew, the king of the gods pointed his magic finger at Arcus and turned him into a little bear. Immediately, Arcus recognized his mother and they embraced.

Unbeknownst to Zeus, Hera had been watching this dramatic scene unfold from afar. Enraged, she charged at Zeus and the bears with fire in her eyes. When Zeus saw her coming, he knew the bears' lives were in danger. To defuse the situation, he reached down and grabbed both bears by the tail. With his godly strength, he swung them around, faster and faster, until, with all his might, he threw the bears into the northern sky, where to this day they are safe from Hera's temper.

Why are the bears' tails so long? Your tail would be stretched out too if someone threw you into the sky by your tail!

CASSIOPEIA THE QUEEN AND CEPHEUS THE KING

The other two major constellations in the northern sky are Cassiopeia and Cepheus. Queen Cassiopeia is easy to find, with its stars as bright as those in the Big Dipper. Many of you have already seen it, even if you didn't know what you were looking at. Just look for a big W in the northern sky. Depending on the time of night and the season, it will either be right side up, upside down, or laying on one of its sides. The W is supposed to outline Cassiopeia tied to her throne. Cepheus looks nothing like a stately king. The constellation looks more like a house with a steep roof. Cepheus is larger than Cassiopeia, but its stars aren't as bright.

Celestial Goodies

M52: This open cluster, variously termed the Scorpion Cluster or the Salt-and-Pepper Cluster, is easy to find with a pair of binoculars or a small telescope.
Magnitude: 6.9

M103: This great open star cluster contains about 160 stars. It was the last object Charles Messier cataloged before his death, though later astronomers used his records to add seven other Messier objects to the list.
Magnitude: 7.0

Delta Cephei: One of the better variable stars, this eclipsing binary star varies in brightness every five and a half days as the two stars rapidly revolve around each other.
Magnitude: 3.5 to 4.4

Erakis: The famous astronomer William Herschel named this reddish star the Garnet Star. It's a long-term variable star, varying in magnitude over a period of about two years.
Magnitude: 3.4 to 4.5

GETTING TIED UP IN THE SKY

How did Cassiopeia and Cepheus wind up in the night sky? It comes down to vanity. According to Greek legend, Queen Cassiopeia and her husband King Cepheus were mortal royalty who ruled over ancient Ethiopia. Cepheus was a mellow guy who spent his time with his buddies fishing, hunting, hanging out at bars, and enjoying life. That was fine with Cassiopeia because she preferred running the kingdom by herself. Anything but mellow, she ran Ethiopia with an iron fist.

Cassiopeia was also extremely vain. She was a beautiful woman but it went right to her head. She was a one of those queens who did the "Mirror, mirror, on the wall" thing every chance she could. Every day, she and her entourage paraded around the kingdom, boasting of her beauty. Peasants, the elite, and everyone in-between listened to this all day long, day after day. Cassiopeia would cry out, "I'm the most beautiful woman in the world!" then grab random passersby and ask, "Am I the most beautiful woman you've ever seen?" Of course, everyone responded YES, YES, YES! Earlier in her reign, a few people actually said no, a crime for which Cassiopeia summoned the ax man in her entourage. The naysayers got their heads cut off, right on the spot. Sucking up was highly encouraged in ancient Ethiopia!

Cassiopeia's ego expanded on a daily basis. Maybe that's why Cepheus hung out at the bar with his buddies instead of at the castle. One day, the queen got carried away. During a walk along the beach, she cried out to Poseidon, the god of the sea, that she was more beautiful than any of his daughters. Furious, Poseidon sent Cetus, a giant sea monster, to destroy the entire kingdom. Cepheus and Cassiopeia almost had to sacrifice their daughter Princess Andromeda to Cetus to avoid the destruction of their kingdom. You can read about that mess in the story of Pegasus the Winged Horse, of Perseus the Hero, and of Andromeda the Princess. Each of these characters has his or her own constellation in the heavens near Cassiopeia.

Despite almost losing her daughter, Cassiopeia remained as vain as ever. One day, she shook her fist

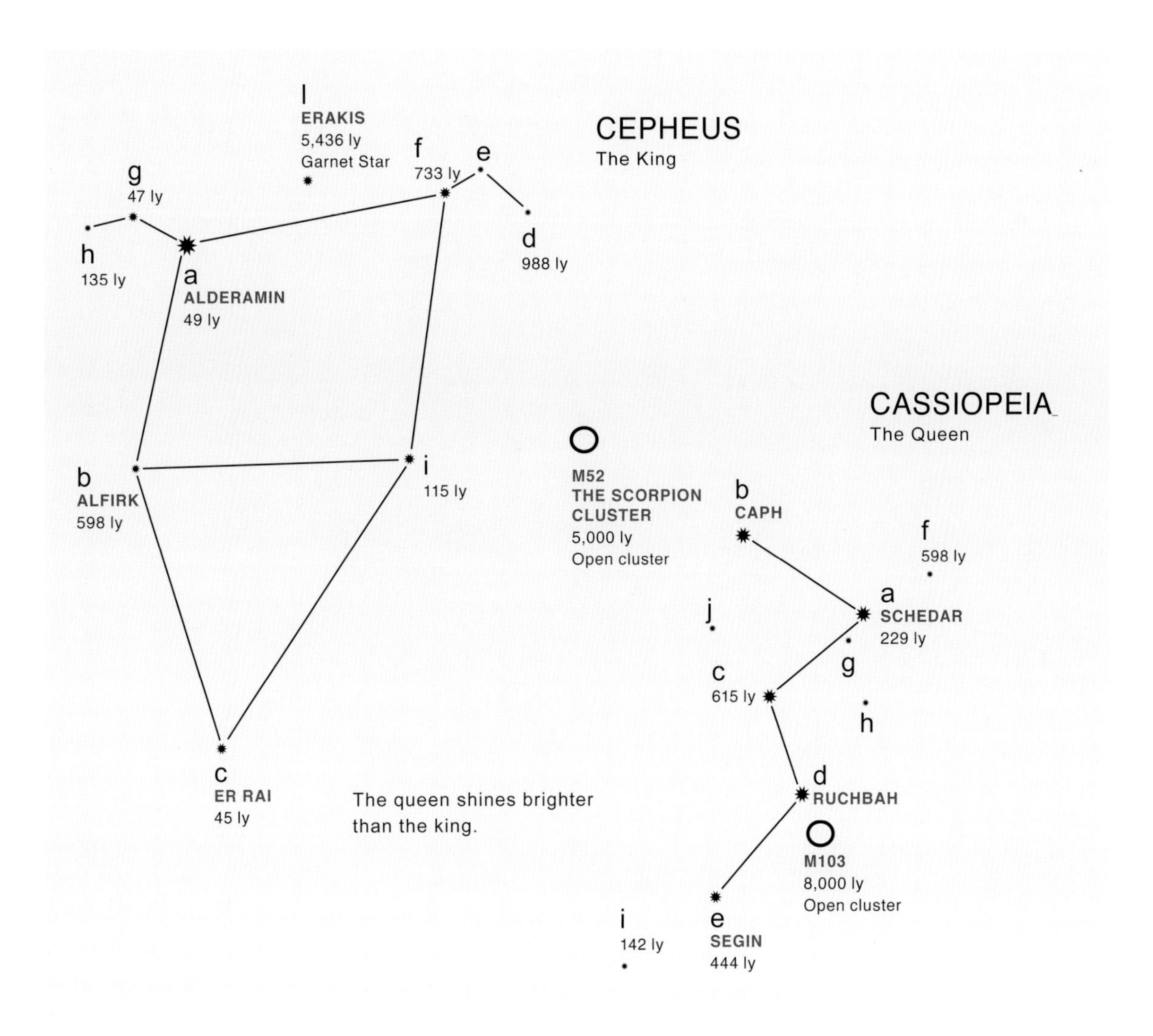

at the sky and proclaimed herself more beautiful than Hera, the queen of the gods. Enraged, the goddess swooped down from the heavens to confront the mortal queen. But Cassiopeia refused to back down and a royal battle ensued, with each queen screaming of her supreme beauty. Finally, Hera stood back and said, "If you think you're so beautiful, I'm going to put you where you can show off your good looks to the whole world every night." With a wave of her arm, Hera tied Cassiopeia to a celestial throne and flung her skyward.

That day, Cepheus came home drunk. He had been partying with the boys at a local pub, but when he heard about Cassiopeia, he sobbed in grief. He cried out to Zeus, the king of the gods, to let him join his wife in the sky. Even though he had spent little time with Cassiopeia, he was still crazy in love with her. At first, Zeus refused, fearing reprisal from Hera, but Cepheus pleaded endlessly. Finally Zeus placed Cepheus in the sky with his beloved wife. To this day—and night—the royal couple circles the northern sky every twenty-four hours, still in love, with Cassiopeia sometimes hanging upside down by her ropes. Talk about a married couple loving for better or for worse!

7

AUTUMN CONSTELLATIONS

While the northern constellations usually hang out in our night sky, the diurnal constellations make only seasonal visits to our latitude. Autumn is great for stargazing because it brings primo constellations like Pegasus the Winged Horse and Andromeda the Princess. Also, the humid summer air—and the mosquitoes—are on the retreat. Drier air means clearer skies. Head outside in September and you'll see what I mean. Plus, darkness falls earlier every night. By November, you can stargaze as early as seven o'clock!

Andromeda Galaxy, M31 (Photograph © Robert Gendler)

PEGASUS THE WINGED HORSE

It's a bird! It's a plane! It's a flying horse! It's Pegasus and he's rescuing Andromeda the Princess. The marquee constellation of fall is definitely Pegasus the Winged Horse. Traditionally this constellation is seen as a horse flying upside down with puny little wings. As much as I like tradition, I must turn away here. I hope I don't make constellation purists angry, but my friends and I see the majestic horse proudly flying right side up, with a huge set of wings. We have to borrow some stars from the constellation Andromeda, but, believe me, it makes a lot of sense.

To best see the great flying horse, face east-southeast any time from September to November, and look for a large square of stars located about halfway between the horizon and overhead. The later in the fall, the higher it will be. Known as the Great Square of Pegasus, these stars make up the horse's torso, even if you insist on seeing him upside down.

In my stargazing world, look for the neck and head of Pegasus emerging from the upper right-hand corner of the square, starting with the star Scheat. These are not the brightest stars, but if you don't stand under a street lamp, you should spot them without too much eyestrain. From Markab, the star at the lower right-hand corner of the square, look for a crooked line of stars outlining the front leg of Pegasus. As you can see, this horse is double-jointed, allowing his leg to bend like it does. That comes in handy for dodging moonbeams.

Off the star Alpheratz, at the upper left corner of the square, you'll see a bright curved line of stars that make up the giant wings of Pegasus. If you look carefully, you'll see a nearly parallel line of faint stars above the great horse's wing. That is Princess Andromeda, catching a ride on Pegasus's wings.

Celestial Goodies

M31: The Andromeda Galaxy appears as a misty patch just above the faint constellation Andromeda. Packed with over a trillion stars, it is the most distant object visible to the naked eye. If your sky is not dark enough, use a pair of binoculars or a small telescope to see this spiral-shaped beauty. We have a semi edge-on view of this galaxy, which stretches over 150,000 light years in diameter.
Magnitude: 4.3

Almach: This beautiful double star consists of one red giant and one greenish-blue star. A small to moderate telescope allows you to easily differentiate them. The stars, separated by more than 90 billion miles, slowly orbit each other as a binary system. With larger telescopes, you can see two other stars in the multiple system.
Magnitude: 2.1

NGC752: Use binoculars or a small telescope to see this very loose open cluster.
Magnitude: 5.7

NGC7662: This planetary nebula glows blue-green, lending it the name Blue Snowball. Use a moderate to large telescope for viewing.
Magnitude: 9.2

M15: This is one of the brighter globular clusters in the sky. In a sphere no greater than 130 light years in diameter, there may be as many as two million stars!
Magnitude: 6.2

M32 and M110: These galaxies are companions to the brighter Andromeda Galaxy.
Magnitude: 10.0 and 10.0

M33: We see this small, local spiral galaxy face-on, unlike our edge-on view of the Andromeda Galaxy. Use a moderate to large telescope to view.
Magnitude: 7.0

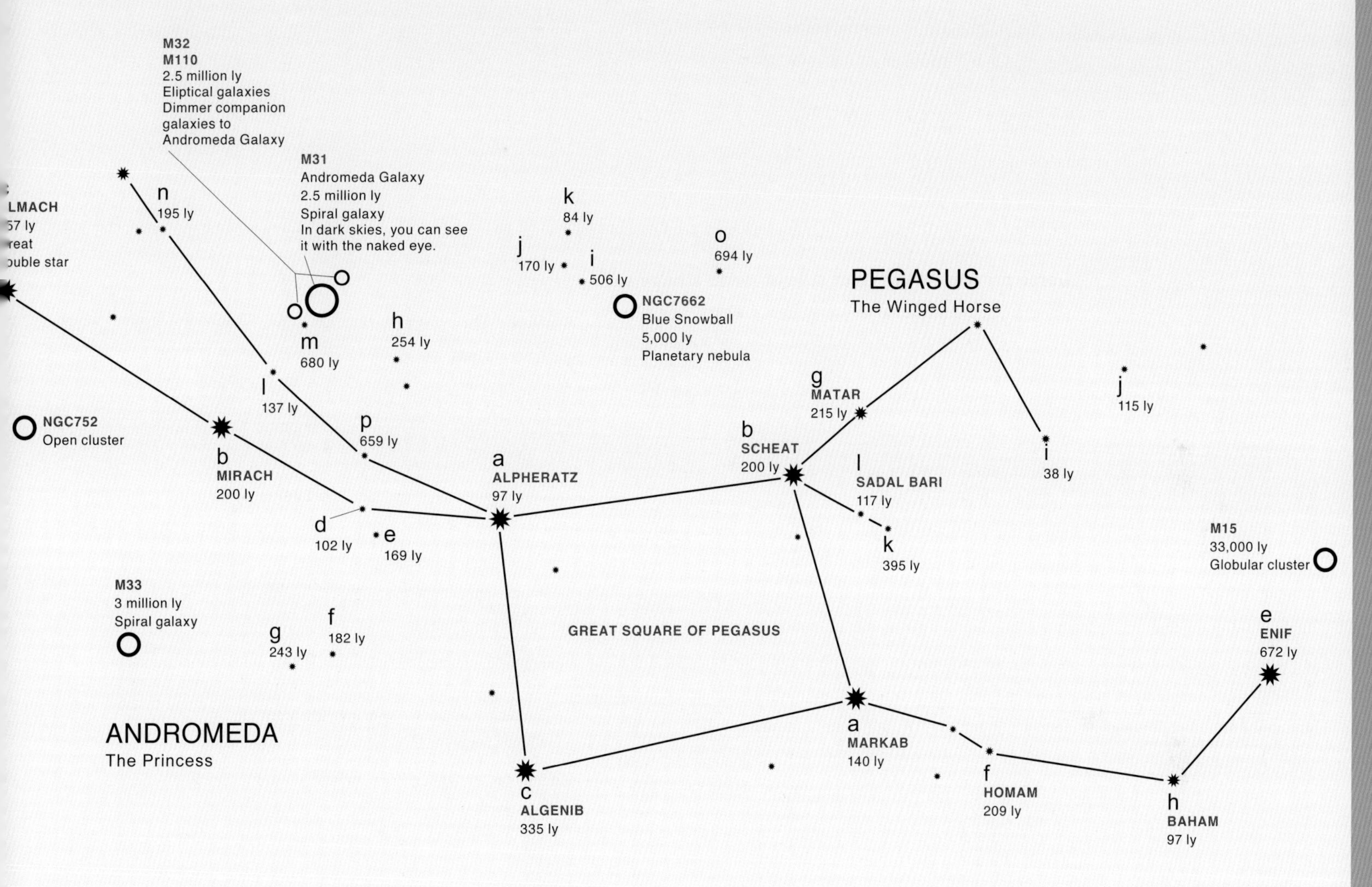
M32
M110
2.5 million ly
Eliptical galaxies
Dimmer companion galaxies to Andromeda Galaxy
M31
Andromeda Galaxy
2.5 million ly
Spiral galaxy
In dark skies, you can see it with the naked eye.
n
195 ly
k
84 ly
j
170 ly
i
506 ly
o
694 ly
PEGASUS
The Winged Horse
NGC7662
Blue Snowball
5,000 ly
Planetary nebula
m
680 ly
h
254 ly
l
137 ly
NGC752
Open cluster
p
659 ly
g
MATAR
215 ly
j
115 ly
b
SCHEAT
200 ly
i
38 ly
b
MIRACH
200 ly
a
ALPHERATZ
97 ly
l
SADAL BARI
117 ly
d
102 ly
e
169 ly
k
395 ly
M15
33,000 ly
Globular cluster
M33
3 million ly
Spiral galaxy
g
243 ly
f
182 ly
GREAT SQUARE OF PEGASUS
e
ENIF
672 ly
ANDROMEDA
The Princess
a
MARKAB
140 ly
c
ALGENIB
335 ly
f
HOMAM
209 ly
h
BAHAM
97 ly

PERSEUS THE HERO

Perseus the Hero is large in lore but faint in the autumn sky. The constellation, hanging above the eastern horizon, is especially tough to spot if your sky suffers from light pollution. Get out to the countryside or at least the outer-ring suburbs. To find Perseus, first look in the high northeast for the bright "W" shape of the constellation Cassiopeia. Perseus is just below Cassiopeia.

Like a lot of constellations, Perseus doesn't look like its namesake. Personally, I see a jet airplane flying north. A kid in my stargazing class once said the constellation looks like the infamously scrawny Charlie Brown Christmas tree. If that's the case, then the star Algol is a bulb hanging from one of the branches. Algol is the second-brightest star in Perseus, but the most fascinating. Known as the Demon Star, Algol almost triples in brightness every three days. An eclipsing variable star, Algol is actually two stars in a tight three-day orbit around each other. When the bigger star covers the smaller one, Algol is at its dimmest. When they are side by side, Algol is at its brightest. You can't resolve the individual stars with a backyard telescope; they are just too close.

Celestial Goodies

NGC869 and NGC884: The famous Perseus Double Cluster is the best double cluster in the sky, visible to the naked eye in dark skies. These clusters are wonderful through binoculars or a small telescope. The two clusters are separated by a couple hundred light years. It's a must see! Breathtaking!
Magnitude: 5.3

M34: This open cluster contains about sixty young stars in an area 5 light years in diameter. It's easily seen with binoculars or a small telescope.
Magnitude: 6.0

Algol: The Demon Star is one of the most famous eclipsing variable stars. Every two days and twenty hours, it drops to a magnitude of 3.4 for about ten hours, before returning to magnitude 2.1.
Magnitude: 2.1 to 3.4

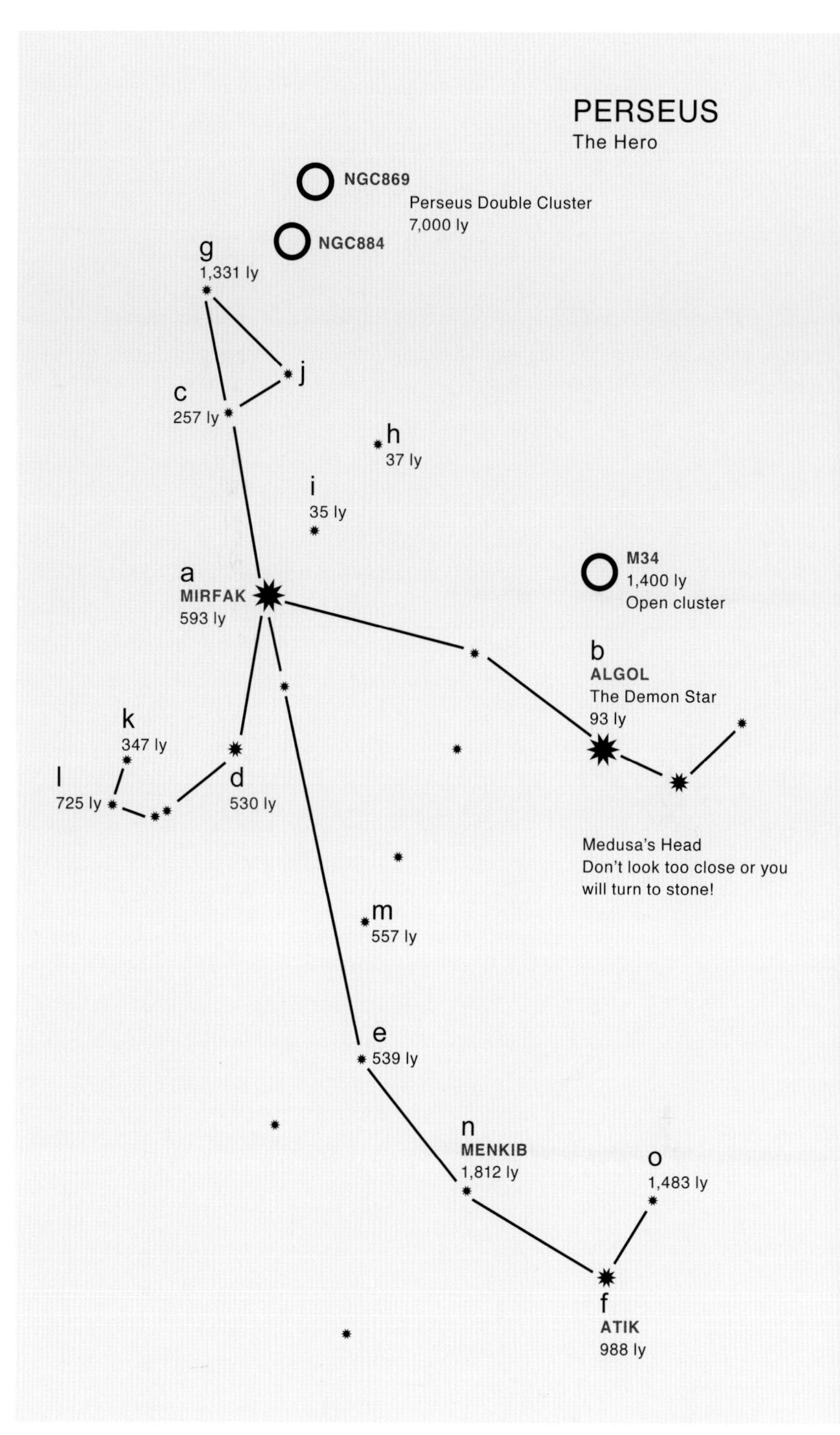

FACING PAGE: Perseus Double Cluster, NGC869 and NGC884 (Photograph © Robert Gendler)

PERSEUS TO THE RESCUE

This celestial soap opera is an offshoot of the saga of the vain queen Cassiopeia and her loyal husband, King Cepheus.

As you recall, Cassiopeia infuriated Poseidon, the god of the sea, when she boasted that she was more beautiful than any of his daughters. Poseidon sent Cetus, his pet sea monster to wreak havoc on Ethiopia. Vanity was about to get Cassiopeia's kingdom destroyed. She and King Cepheus had to do something, and fast! The royal couple consulted a wise oracle, who advised them to chain their daughter, Princess Andromeda, to a boulder on the beach as a sacrifice to Cetus. That would satisfy the sea monster enough to prevent him from wiping out the kingdom. That afternoon, the king and queen took Andromeda out for what she thought was a day at the beach. When the family arrived on the sandy shores, Cassiopeia and Cepheus quickly chained their daughter's arms to a boulder, then ran off. What a mom and dad!

When Andromeda saw Cetus coming, she futilely tried to rattle out of her chains. The situation looked bleak for Andromeda, but happily, her hero was on his way. Perseus, the son of Zeus, was flying back from a mission wearing the winged shoes of Mercury. He was feeling good because he had saved a lot of lives. He had killed Medusa, a gorgon so ugly that if you glanced at her for a fleeting moment, you would turn to stone. Instead of hair coming out of her head, there were live snakes. Using a magic shield, Perseus managed to cut off Medusa's head without looking at it.

Perseus was flying overhead with Medusa's head in a large garbage bag when he saw the beautiful princess and the sea monster closing in on her. Our hero grabbed the gorgon's head out of the garbage bag and shook it at Cetus. The sea monster immediately turned to stone and sank into the ocean like a boat anchor.

Here's where the story gets really bizarre. The blood dripping from Medusa's severed head hit the ocean and magically produced a winged horse that answered Perseus's every command. At his bidding, the winged horse flew to the beach, chewed off the chain that held the princess, and brought her back to the hero.

Perseus married Andromeda and named his special new horse Pegasus. The hero enjoyed many years of wedded happiness, despite having the vain Cassiopeia as a mother-in-law. Then Perseus met his demise. He and his buddies were celebrating in a local bar after a night of successful demolition chariot races. They had consumed several rounds of brew, when a gang of thugs came into the bar and started a sword fight. Perseus and his comrades beat back the thugs, but in the process Perseus took a sword in his neck.

Zeus and the rest of the gods on Mount Olympus decided to honor Perseus by placing him in the sky as a constellation. When Andromeda died a few years later, the gods put her in the celestial dome as well, not far from her husband, riding on the wings of Pegasus. Together they sweep across the skies with Cepheus and Cassiopeia, still in love. By the way, the variable star Algol is located where our hero holds the head of Medusa. Don't look too closely; I don't want you turned to stone!

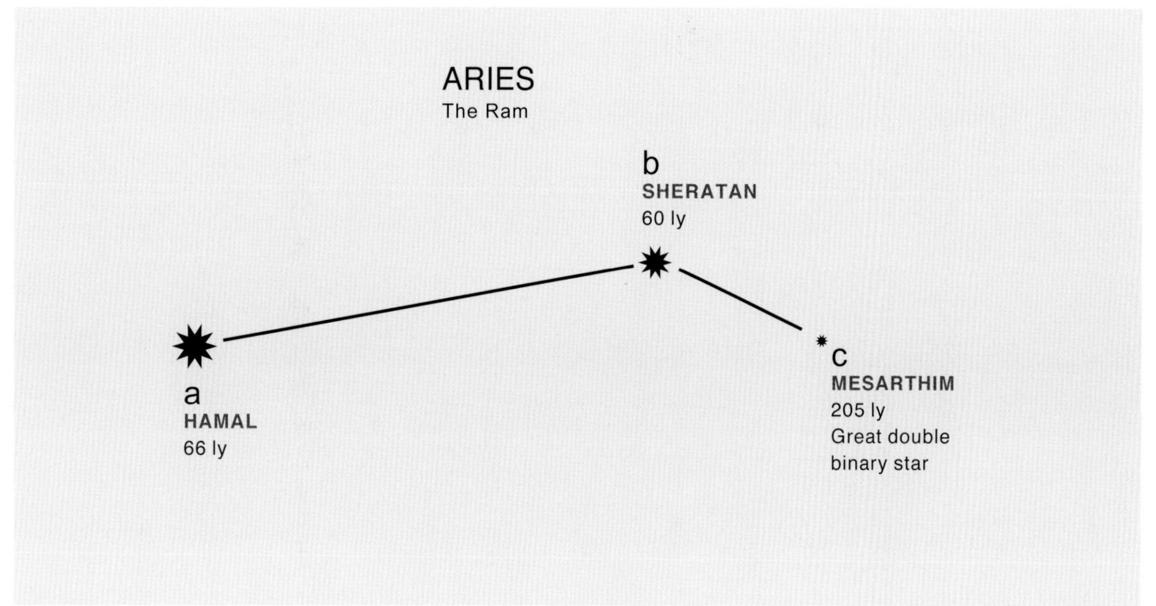

ARIES THE RAM

Aries the Ram is a small constellation with a big story. Aries comprises just three main stars: two moderately bright stars and a dim star, all of which represent the horn of the ram. Look for Aries in the eastern autumn skies, below the wings of Pegasus. The constellation actually occupies a larger area of the sky than the little horn we see, but the rest of it is faint and undefined. The two brighter stars in the horn are Hamal and Sheratan, and the dimmer star to the lower right is Mesarthim. Hamal is a giant star over 66 light years from Earth. It's 37 times larger than our own sun and over 426 times as luminous.

Once upon a time, Aries was the backdrop constellation for the sun during the first day of spring. Because of the wobble in the earth's axis, Aries has been replaced in the background during the vernal equinox by the faint constellation Pisces the Fish.

Celestial Goodies

Mesarthim: This great double binary star is easily split with a small to moderate telescope. The two stars are separated by 36 billion miles, and revolve around each other every 3,000 years. Magnitude: 3.9

THE LITTLE RAM WITH A BIG STORY

The Greek mythological story of Aries the Ram is a sweet one. It reminds me of the old TV show *Lassie*. Zeus, the king of gods, had a pet ram named Aries. He was a grand ram with a coat of golden fleece, and wings to soar the skies above Mount Olympus.

One day Zeus and one of his many girlfriends were picnicking in a lush valley at the foot of Olympus, when Apollo, the god of the sun, shouted from high in the sky. Ten miles away, a lion was bearing down on two small children. The kids had slipped away from their mother at a marketplace and were in some nearby brush. Zeus was in a good mood that day, and he knew his pet ram loved kids, so Zeus pointed Aries in the right direction and sent him on a rescue mission.

The lion was within 20 feet of bagging the kids when Aries swooped from the sky like a cruise missile. He scooped up the children on his back and flew to the marketplace, where the kids were reunited with their relieved mother.

For the rest of his life, Aries performed missions of mercy. When Aries died, Zeus rewarded his bravery by placing his body in the heavens to become the constellation we see today.

8

WINTER CONSTELLATIONS

For stargazers, these are the best of times and the worst of times. Stargazing may not seem user friendly in the icy cold of winter, but this is when the air is driest and clearest–and the constellations are the best and brightest! So bundle up, fill a thermos with hot cocoa, and stay outside as long as you can bear. Just be sure to let your telescope cool off outside for at least forty-five minutes. Otherwise the lens and mirrors will get out of whack. Take a little walk. It will warm you up and ease the shock of the winter cold.

One of the treasures of the winter sky is the Winter Triangle. It's made up of three bright stars: Betelgeuse in Orion the Hunter, Sirius in Canis Major, and Procyon in Canis Minor. They form a perfect triangle. Go out and see it!

So many bright stars cram the winter night sky that you would think we're facing the center of our galaxy, but we're actually facing 180 degrees the other way, toward the edge of our galaxy.

Orion Nebula, M42 (Photograph © Robert Gendler)

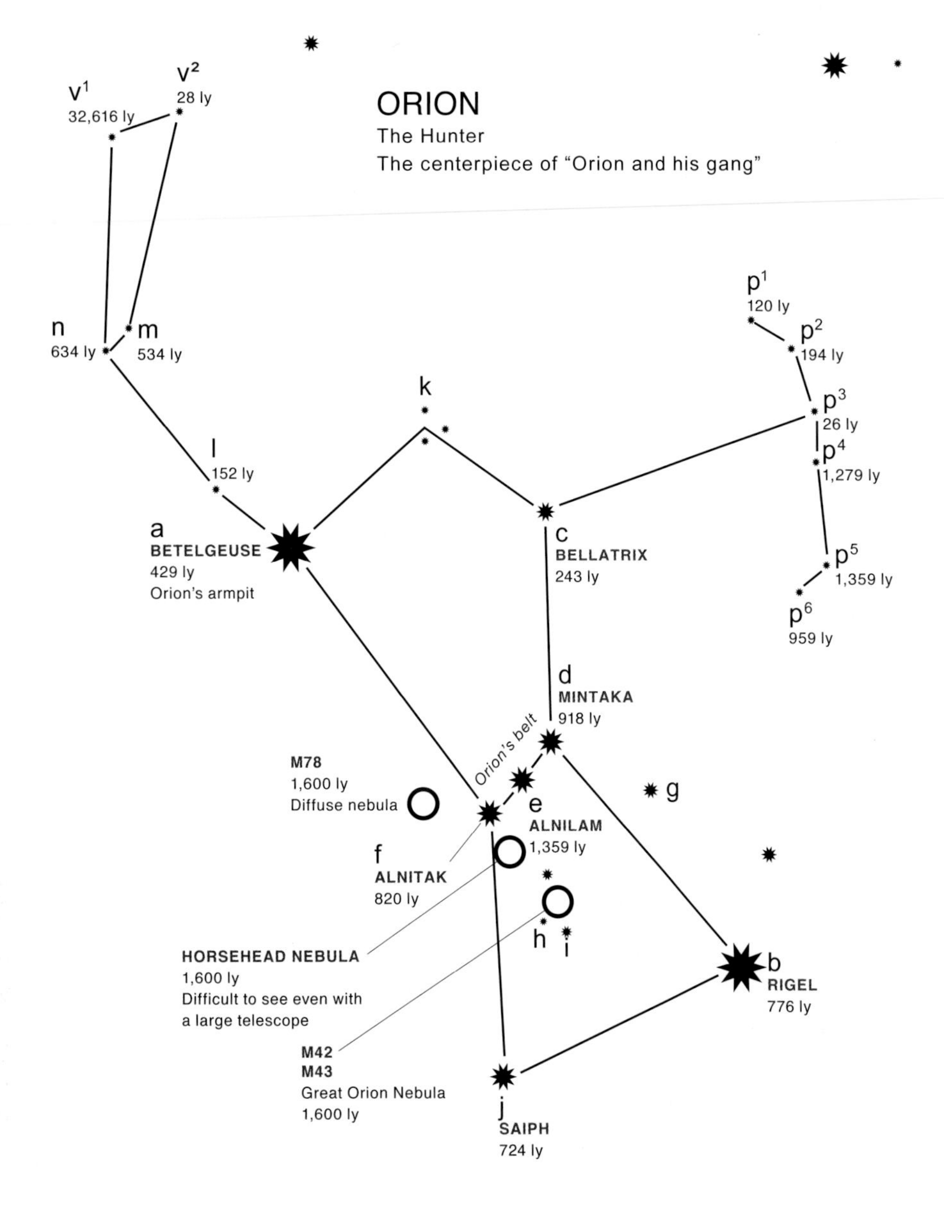

ORION THE HUNTER

Orion the Hunter serves as the hub of the winter constellations and, along with the Big Dipper, is probably the most recognizable star pattern. It's one of the few constellations that almost looks like its namesake. Most people can imagine Orion as the figure of a giant person, though some stargazers see a bowtie on its side and I see an hourglass. The brightest part of the constellation is the torso of the great hunter. Two fainter lines outline Orion's arms. In one hand he holds a club and in the other, a shield. In less politically correct interpretations, like mine, Orion holds a lion by the scruff of its neck.

Orion's hallmark is his belt, made of three bright stars in a row. Nowhere else in the night sky, from anywhere in the world, will you find three bright stars lined up quite like these. From the lower left to the upper right they are Alnitak, Alnilam, and Mintaka. Despite their precise alignment, these stars are not physically related. They are hundreds of light years apart and their arrangement in our sky is coincidental.

The brightest star in Orion is Rigel, which marks the hunter's left knee. It's the fifth-brightest star in our night skies, over 770 light years away and over 50,000 times more luminous than the sun. If Rigel were a closer star—say as close as the star Proxima Centauri, the closest nighttime star to the earth, at a distance of just over 4 light years—it would shine as bright as a full moon and cast shadows. Somehow, I don't think anyone would write songs titled "Harvest Rigel" or "Rigel over Miami" though.

The second-brightest star in Orion is Betelgeuse (pronounced beetle-juice), just under 430 light years away. Betelgeuse sits at the armpit of the hunter and, in fact, Betelgeuse means "armpit" in Arabic. The next time you see Betelgeuse, keep in mind that it's the biggest single thing you've ever seen! This red giant fluctuates in diameter from over 313 million miles to around 1 billion miles. It's the largest star within 1,000 light years of the earth. If you were to replace our sun with Betelgeuse at its largest, this gigantic star would encompass the planets Mercury, Venus, Earth, Mars, and even Jupiter! Some astronomers estimate that within a million years or so, this star will explode in a supernova. That would be quite a show, but I'm not waiting up for it.

Below the belt of Orion is M42. It's a fuzzy, ghostly looking patch of light in the middle of Orion's sword. What you're seeing is the Great Orion Nebula, part of a huge cloud of hydrogen about 20 to 25 light years across and around 1,600 light years away, lit up by the young stars within it. With even a small telescope, the nebula casts a greenish glow. Four stars right in the middle light up the nebula like a florescent bulb. These are hot young stars, possibly only 10 to 20 million years old and maybe even younger, which would make them stellar infants. Radiation pouring out of these young bucks continuously ionizes their hydrogen birth cloud, giving it the glow we see from Earth.

Celestial Goodies

M42: The Great Orion Nebula is the best emission nebula in the night sky. You can actually see four young stars in this baby star factory arranged in a trapezoid known as the Trapezium. You'll need a good pair of binoculars or a small telescope.
Magnitude: 5.0

M43: This nebula is actually a part of the Great Orion Nebula, but a dark area makes the objects look separate. You'll need a small to moderate telescope.
Magnitude: 7.0

Horsehead Nebula: This nebula is one of the hardest to see but worth the effort. A dark nebula in front of a luminous one creates the shape of a horse's head. You need a super-dark sky, a large telescope, and a lot of patience!
Magnitude: 10.0+

M78: This faint, diffuse nebula is actually part of the large cloud that makes up the Great Orion Nebula.
Magnitude: 8.0

CANIS MAJOR AND CANIS MINOR

These dogs of the night sky are part of "Orion and his gang." Like a lot of constellations, they are often referred to by their traditional Latin names. Canis is Latin for dog, and Canis Major and Canis Minor mean Big Dog and Little Dog, respectively.

These hounds of winter are easy to spot. Canis Major stands up on his hind legs to the lower left of Orion. The brightest star in the entire night sky, Sirius perches on the Big Dog's nose. Just extend a line to the lower left from the three bright belt-stars of Orion and you'll run into Sirius.

The proximity of Sirius to the earth, only 8.6 light years, accounts for its brilliance in our night sky. Compared to a lot of other stars, Sirius is wimpy, only about 2 million miles in diameter and about twice as big as the sun. Many stars in our part of the galaxy are much larger. Remember that the brightness of a star is determined by size and distance. Just because a star is dim doesn't mean it's puny. It could be a monster many light years away.

To find the rest of Canis Major, look for a star to the right of Sirius. That's the front paw of the Big Dog. Go back to Sirius and look to the lower left for a distinct triangle of stars that outline the dog's rear end, hind leg, and tail. The star at the tip of tail, Aludra, is over 3,200 light years away. The light that you see from Aludra left that star in the year 1200 B.C.!

Canis Minor is a poor excuse for a constellation. All there is to the Little Dog is a bright star called Procyon and a couple dimmer stars just above it. It's a Chihuahua! Look for Canis Minor just to the left of Orion.

One of the great treasures of the winter sky is the Winter Triangle. It's made of three stars: Sirius in Canis Major, Procyon in Canis Minor, and Betelgeuse in Orion. These stars form a perfect triangle. Go out and see it!

Celestial Goodies

M41: This open cluster is 20 light years across, with a reddish star shining right in its middle.
Magnitude: 5.0

THE RISE AND FALL OF THE GODFATHER OF THE SKY

Just as the constellation Orion is rich with astronomical treasures, the myth of the mighty hunter is rich with lore. Many different cultures have their own story of this ancient constellation. My favorite tale trickles down from Greek and Roman mythology and involves Artemis, the goddess of the moon.

Orion the Hunter was a man's man. He was big and strong, and he loved to hunt and fish. Like most of the animals he hunted, Orion was nocturnal. He stalked his prey by night and slept under a giant tree by day. He was a shy giant who didn't like to mix with other people or the local DNR, which took exception to Orion's nocturnal hunting practices. So he moved to a large deserted island, where he could hunt and fish undisturbed.

Life was great for the reclusive hunter. Every night he and his hunting dogs were out slaying beasts of all kinds. He also had a secret admirer, Artemis, the great moon goddess. Every night as Artemis flew across the sky, guiding her magic moon chariot pulled by flying horses, she longed to be with the mighty hunter. But she could hardly jump down and join Orion. Not only would she be ignoring her

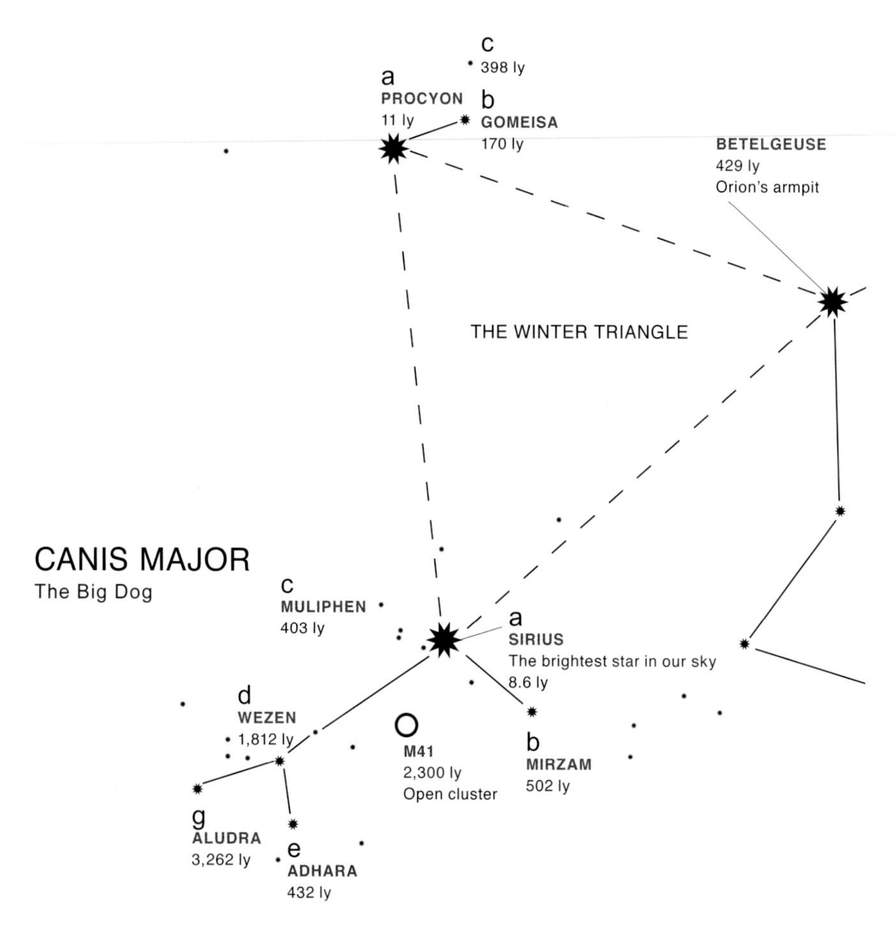

duties as moon goddess, but she would also be mixing with a mortal and that would get her in trouble with her father, Zeus, the king of the gods!

One night as Orion was cleaning up on a herd of wild boars, Artemis couldn't take it anymore. She yelled whoa to the horses and headed down to Orion's island. Finally, she met him eye-to-eye. It was love at first sight. Orion gave up the hermit life after one look at the goddess. Right away, she changed out of her royal robes into hunter's blaze-orange, and they hunted through the night. But when dawn approached, Artemis jumped onto her moon chariot and raced to the horizon. The next night, she halted the moon mid-sky again and joined her new love for another night of hunting.

This love affair went on night after night, until Zeus learned about his daughter's behavior through the godly grapevine. He had to stop the affair, but without losing his daughter's love. So Zeus came up with a plan to have Orion killed and make it look like an accident. Zeus made arrangements for a giant scorpion to be dropped on Orion's island during the day, while the hunter slept.

The day of the assassination arrived. Orion fell into a deep sleep, happy from a successful night of hunting with his divine girlfriend. Soon the scorpion crawled into his camp and approached the peaceful hunter. A loud cry broke the air. It was Orion's security system, his friend the mockingbird. Orion bolted awake just as the scorpion attacked. What followed was a battle that went on for hours. Finally, the hunter locked the scorpion's head under his arm and prepared to break its neck. Just then, Artemis rose with the moon in the eastern sky. Orion glanced up, and the scorpion lunged out of his hold and delivered the fatal sting to Orion's neck.

Artemis raced to the scene, but she was too late. Orion was dead. As the scorpion tried to retreat, the moon goddess grabbed it by the tail and flung it so far that it landed in the sky, where it became the constellation Scorpius. Artemis returned to the slain Orion and wept for hours. Finally, cradling his body in her arms, she flew into the sky. When she was high enough, she gently tossed Orion in the sky, turning him into a bright constellation. This way, he would still be with her every night. She also placed Orion's two favorite hunting dogs, Canis Major and Canis Minor, next to him.

Artemis made sure that Orion was on the opposite side of the sky from the scorpion that assassinated him. That's why we never see the constellations Orion and Scorpius in the sky at the same time. As soon as Orion rises in the east, the scorpion sets in the west.

AURIGA THE CHARIOTEER

One of the strangest constellations in the winter heavens is Auriga, the chariot driver who is hauling a mama goat and her kids on his shoulder. I know the Greeks went without books or movies and had to use the constellations to pass their stories from generation to generation, but they went to extremes with this one! It must take an amazing imagination, and perhaps a few libations, to see this detailed scene in a simple pentagon-shaped constellation.

Auriga, the northernmost constellation in "Orion and his gang," perches just above the hunter's head. It is shaped like a giant lopsided pentagon, with the bright star Capella marking one corner.

Capella is the brightest star in Auriga and the fourth-brightest star in our night sky. Latin for "she goat," Capella is where the mama goat sits on the chariot driver's shoulder. Look right next to Capella for the dim triangle of stars that make the kids. The closest kid star to Capella is called Almaaz and it's no kid. The general consensus is that Almaaz is a little over 2,000 light years away, which means the light we see in the night sky left that star more than two milleniums ago. Almaaz is more than 140 million miles in diameter, and over 40,000 times as luminous as the sun.

Celestial Goodies

M36: Sixty to seventy young stars, about 30 million years old, make up this bright open cluster. It is about 15 light years across and viewable through a small to moderate telescope. Magnitude: 6.5

M37: This is another bright open cluster of several hundred stars in an area about 25 light years across. A bright orange star shines in the middle. Use a small to moderate telescope. Magnitude: 6.0

M38: Yet another large open cluster, this one is over 220 million years old, which is old for an open cluster. Call it a teenage cluster! You can easily see it with a small to moderate telescope. Magnitude: 7.0

THE GREAT CHARIOT RACE

According to Greek legend, there once was a mighty king named Oenomaus who ruled a mighty kingdom. He had a beautiful daughter named Hippodameia, who had many suitors. Oenomaus didn't want his daughter to marry any of them and, in fact, he wanted them all killed. The king was an excellent chariot racer, so he arranged races with each suitor. The first to beat Oenomaus would win the hand of his daughter, but if the suitor lost the race he would be killed. Oenomaus had the fastest horses in the land, so he easily beat all the young opponents and killed the losers one by one.

Only one suitor remained. Pelops was the son of Hermes, the messenger of the gods. When his turn came, Pelops got some divine help from the gods. They crafted a chariot that sprouted golden wings. Pelops didn't stop there, though; he convinced the king's chariot driver, Myrtilus, to replace the chariot lynchpins with wax copies. In exchange, Pelops promised half the kingdom to Myrtilus if the king lost the race and was killed.

When the race began, Oenomaus kept up with Pelops, but as soon as the golden wings popped out of the crooked suitor's chariot, the king was left in a cloud of dust. When Oenomaus ordered his driver to spur the horses on, Myrtilus leapt out and the chariot fell apart. Oenomaus was dragged to his death, cursing the name of his driver.

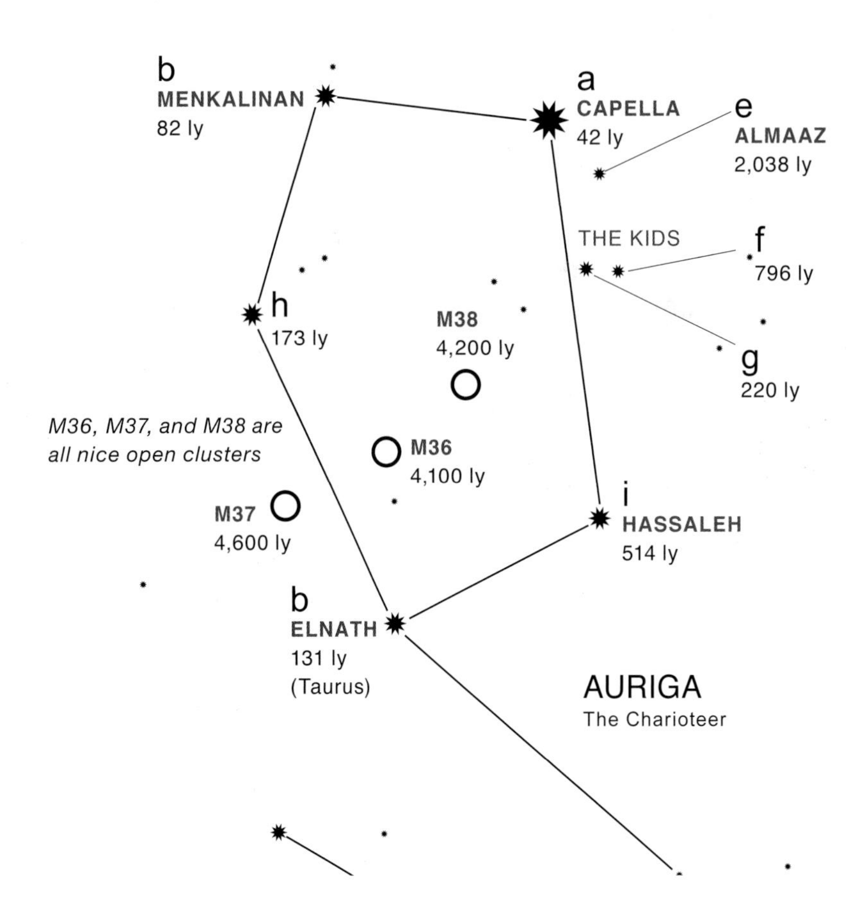

WINTER CONSTELLATIONS

Pleiades, M45 (Photograph © Russell Croman)

After Pelops and Hippodameia married, Myrtilus showed up demanding his half of the kingdom. A deal was a deal! Pelops, crook that he was, told Myrtilus that his lawyers were drawing up the papers and they would be ready in a few days. Satisfied, Myrtilus turned to walk off. Pelops, with his inherited godly powers, kicked Myrtilus so hard that he flew into the sky and became the constellation we know as Auriga.

No one knows exactly how the betraying chariot driver got the mama goat and baby goats on his shoulder, but the leading theory is that shepherds added them as they watched their flocks by night.

TAURUS THE BULL

Taurus the Bull is a small but distinct constellation in "Orion and his gang." The best way to find it is to first look for the Pleiades, a bright star cluster that jumps out at you in the eastern heavens. This cluster looks like a tiny Little Dipper, and some people mistake it for the actual constellation. The Pleiades, otherwise known as the "Seven Sisters," is a large cluster of hundreds of young stars in an area 30 light years across. These stars are about 80 to 100 million years old and over 400 light years from our backyards.

Most people only see six stars in the Pleiades with their naked eye, even in the darkest conditions. Why, then, is the Pleiades called the Seven Sisters? No one knows for sure, but some believe that since this cluster was first named thousands of years ago, one of the stars has faded.

The Pleiades are also known as the Halloween Cluster, because it's nearly overhead at midnight on October 31. In many ancient cultures, this cluster was believed to foretell great disasters or, possibly, the end of the world. Some European cultures saw the cluster as the collective light from souls of people who had died in the last year.

The Japanese name for the Pleiades is Subaru. Sound familiar? It is. The Subaru Automobile Corporation, formed in the early 1950s from six smaller auto companies, named itself after the brightest star cluster in the sky. In fact, their logo looks like a star cluster. Actually, the old logo, discontinued several years back, looks more like the Pleiades.

Once you've spotted the Pleiades, look below the cluster for a small, dim-but-distinct arrow pointing to the right. A loose open cluster called the Hyades, this arrow outlines the snout of Taurus the Bull. On the arrow's lower side is a medium-bright reddish star, Aldebaran, which marks the bull's ruddy eye. Each side of the arrow extends to the far left, forming the beast's horns. One horn ends at Elnath, which is technically part of Taurus but is also one of the stars in the constellation Auriga the Charioteer.

TAURUS

The Bull

The main part of Taurus is the small V, or arrow, outlining the bull's snout.

b
ELNATH
131 ly
Also in Auriga

M1
Crab Nebula
Remnant of a supernova that exploded in 1054 AD
6,500 ly

M45
Pleiades Star Cluster
Seven Sisters
410 ly

f
418 ly

e
155 ly

d
146 ly

c
154 ly

a
ALDEBARAN
65 ly

k
370 ly

Celestial Goodies

M45: The Pleiades, the best open cluster in the sky, resembles a mini Little Dipper. It covers an area over 4 moon-lengths in size. You can best see it with binoculars because it's so big. Telescopes ruin the view!
Magnitude: 1.4

M1: The Crab Nebula is the ghostly remnant of a supernova, a star that exploded in A.D. 1054. The fresh supernova was visible in the daytime for twenty-three days after the explosion and visible to the naked eye in the night sky for another two years. Since 1054, it has faded, but it is still visible with a small to moderate telescope. Material from this explosion has expanded to a diameter of 40 trillion miles and still moves out in space at over 600 miles a second! Within the Crab Nebula is a spinning neutron star, otherwise known as a pulsar.
Magnitude: 9.0

THIS STORY'S A LOT OF BULL!

The tale of Taurus the Bull is one of deception. Zeus, resident playboy of Greek mythology, used all the tools he had to lure the ladies. One of his love targets was Princess Europa, the daughter of a Phoenician king. Zeus first met Europa at a local holiday party, but she was under-whelmed by him. He even got her under some mistletoe and nothing happened.

Zeus had to get creative to win Europa. He knew that, as a hobby, the princess loved to raise prize bulls. She would spend hours in the pasture with her beautiful beasts. Being the king of the gods, with all kinds of magical powers, Zeus turned himself into Taurus, a gorgeous white bull with golden horns, and wandered into Europa's pasture. The princess was delighted by Taurus's beauty and tameness, and spent hours grooming the god-in-bull's-clothing.

Europa felt so at ease with Taurus that, one day, she put a saddle on his back and hopped on. This was the opportunity Zeus was waiting for. After a few gentle rides around the pasture, he kicked into gear and shot across the countryside. Europa was frightened but also excited by the adventure. Taurus reached the sea but that didn't stop him. He charged into the waves with Europa clinging to his back. He swam all the way to the Greek Island of Crete and finally stopped (Europa was wet and sunburned when she rolled off the deceptive bull). It was then that Zeus revealed his true identity.

Somehow, that did it! Europa fell in love with Zeus and they were an item for a couple of years. Zeus, however, was not into long-term relationships and was on the verge of dumping Europa, when the princess beat him to it. After a long day of ruling the heavens, Zeus came home to find the door locks changed and his godly clothes in the front yard. Even though he wasn't a bull anymore, Europa still put Zeus out to pasture!

Zeus never forgot the beautiful princess who dumped him. Years later, he decided to honor her memory by placing the shape of the bull in the evening sky.

GEMINI THE TWINS

Of the sixty-six constellations we see throughout the year, Gemini is one of the brightest. It is located west of Orion, making it the last major winter constellation to linger in our early evening spring skies.

Finding Gemini is easy in the cold, clear winter skies. Just look for two identically bright stars, right next to each other, to the upper left of Orion. These stars are Castor and Pollux, less than 5 degrees apart, which is equivalent to half of one fist-width. Castor and Pollux mark the heads of the twins. If it's dark enough where you're stargazing, you'll see two faint parallel lines of stars to the lower right of Castor and Pollux. These lines make up the bodies of the twins and remind me of stickmen that kids like to draw. The feet of Castor and Pollux are not far away from Betelgeuse, Orion the Hunter's armpit.

If you slowly scan the constellations with binoculars or a telescope, you'll see a lot of open star clusters that are young stars born out of the same nebulae. Castor and Pollux, though, are a story all by themselves. Pollux is a giant star, 34 light years away and more than 9 million miles in diameter, eleven times the diameter of the sun. Pollux sports a surface temperature of over 8,000°F, a little cooler

Crab Nebula, M1 (Photograph © Russell Croman)

than our home star. Castor looks like a single star to the naked eye, but a moderate telescope reveals a beautiful double star, and larger telescopes reveal a bizarre collection of six stars revolving around each other in an intricate cosmic ballet.

Celestial Goodies

M35: This large, bright open cluster can easily be seen with binoculars or a small telescope. Magnitude: 5.5

NGC2392: This faint planetary nebula, nearly 4 trillion miles across, was named the Eskimo Nebula for its shape, which resembles an Eskimo's hooded head. It also reminds me of the character Kenny on the TV show *South Park*. Some astronomers think it's expanding nearly 2 billion miles a year. You'll need a moderate to large telescope to see it well. Magnitude: 9.9

THE BEST OF BROTHERS AND THE BEST OF FRIENDS

According to Greek mythology, Castor and Pollux were the twin sons of Leda, the mortal queen of Sparta. The twins, incredible as it sounds, had different fathers. Castor was the son of Leda's husband, King Tyndarus, while Pollux was the son of Zeus, the king of the gods. How can I put this? Let's just say that nine months before Castor and Pollux were born, Queen Leda had quite a night! Greek mythology is full of this kind of behavior.

As a result, Castor was a mortal and Pollux was a half-god. The twins grew up together in the castle and had the finest of everything, including a great education. Castor became one of the finest horsemen in the land and Pollux became a champion boxer.

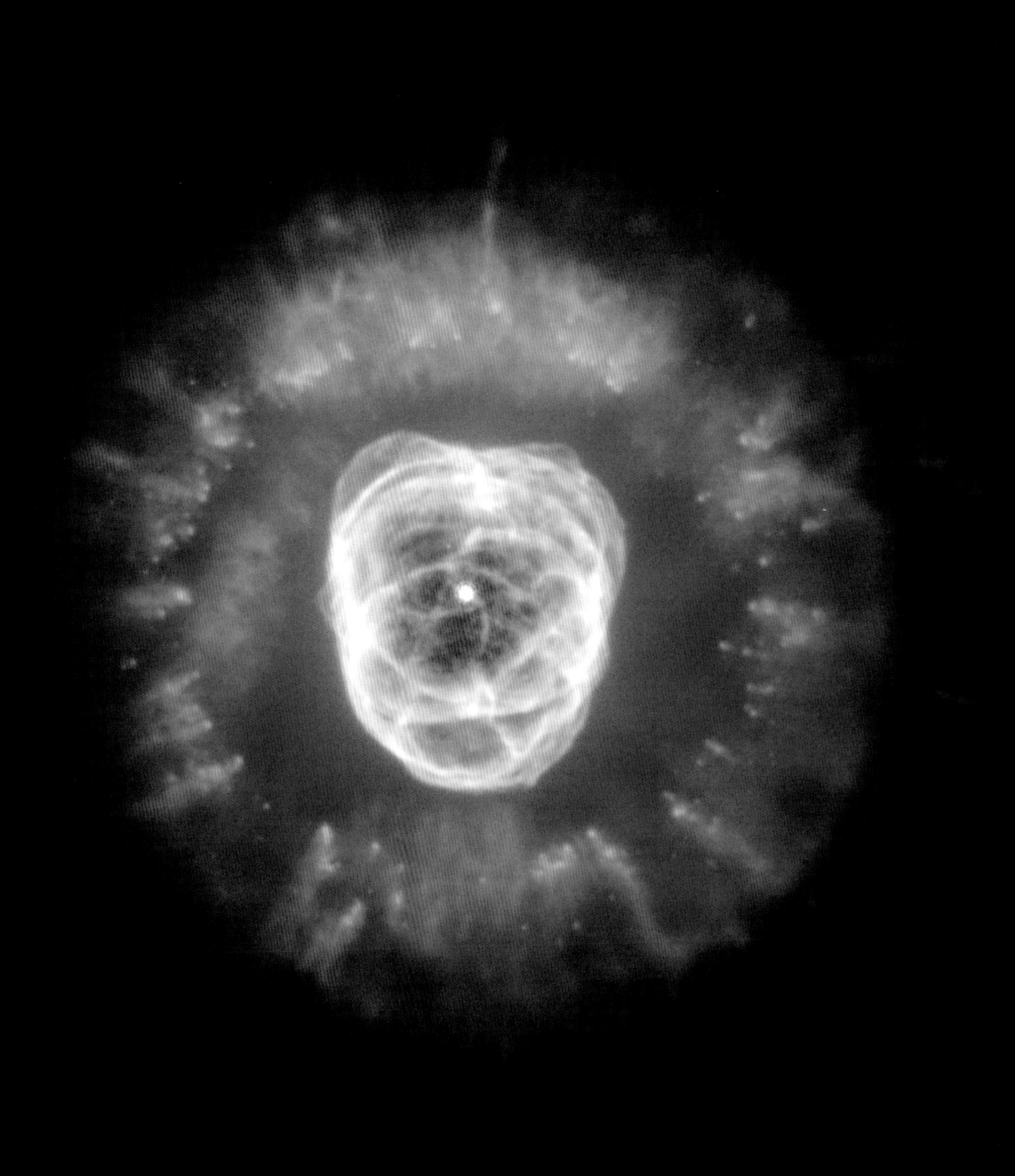

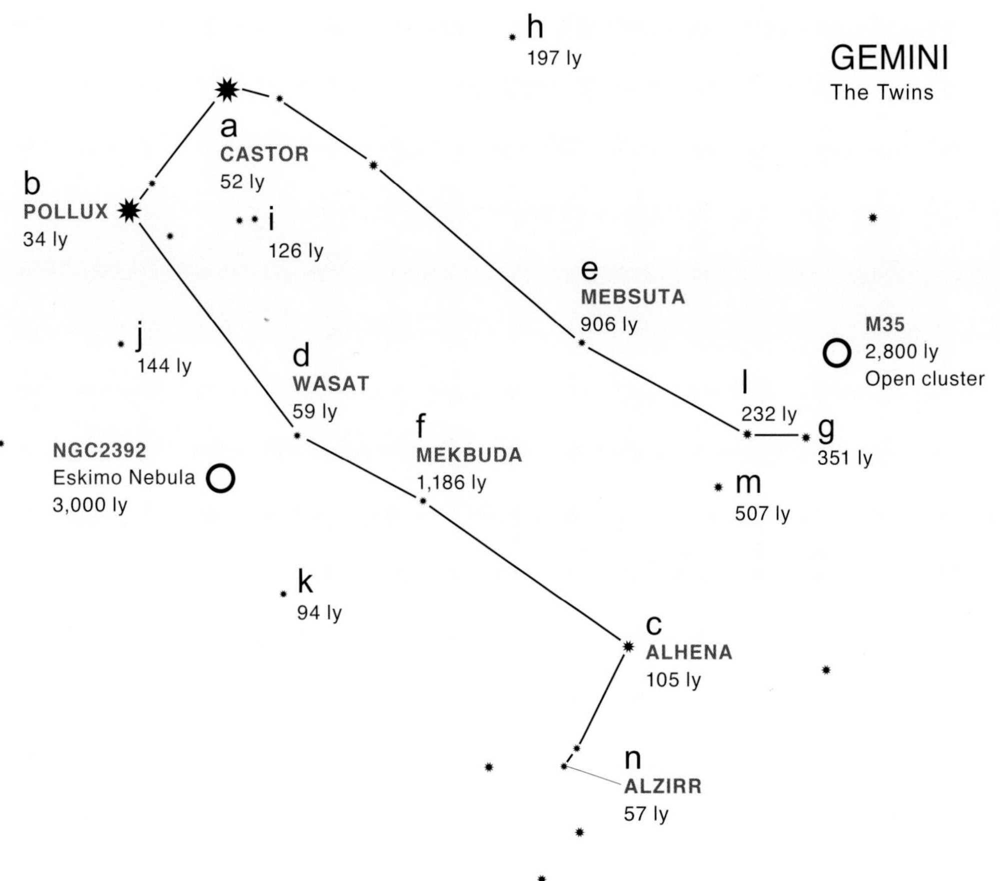

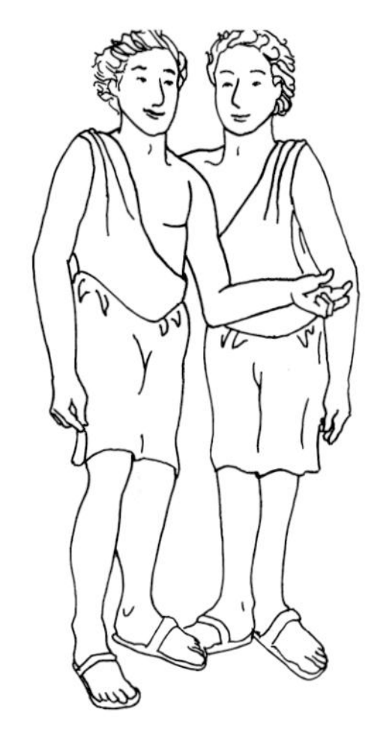

Castor and Pollux were best friends and often went on trips together. One evening, the vacationing twins stopped in a bar for drinks and met two beautiful young women. The twins bought the ladies drinks, and soon all four were dancing. Then two local men walked into the bar. They were big and had the ill-tempered look of just having left a long day's work. It turned out they took exception to their girlfriends dancing with a pair of glitzy strangers. A sword fight broke out (in those days, that's how scores were settled). Castor and Pollux took the duel outside and gave the local boys a real fight, eventually slaying them. During the melee, Castor took a sword in the side, but it didn't seem serious. Unexpectedly, the gash became infected and turned into a mortal wound, and Pollux lost his brother and best friend.

Castor went to the underworld, and Pollux missed him like crazy. He longed to join his brother, but his godly blood made him immortal and he would never be allowed in the underworld. Pollux begged and begged his father Zeus to allow him to see his brother again. At first, Zeus declined, but he finally arranged to have Pollux spend half of each day in the underworld with his brother.

The relationship between Castor and Pollux was so special that Zeus had their figures etched in the sky as the constellation we see today. Early Greek and European sailors saw this constellation as a good luck charm. When they saw Gemini rise before sunrise, it was a sign that the winter and spring storms were over and the summer sailing season had begun.

FACING PAGE: Eskimo Nebula, NGC2392 (Courtesy of NASA, A. Fruchter, ERO Team, STScI)

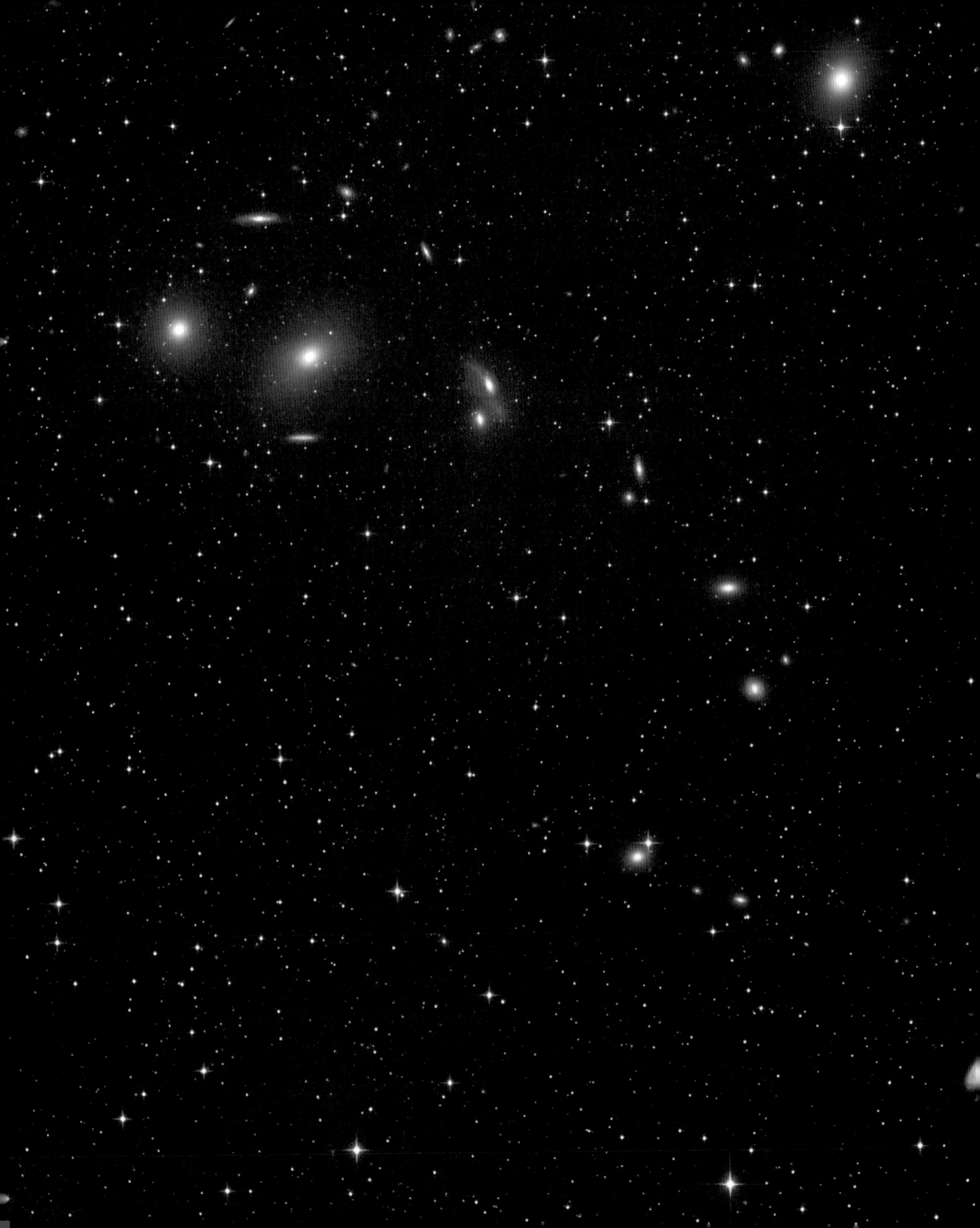

9

SPRING CONSTELLATIONS

The nights grow warmer as the spring constellations rise in the east. Spring stargazing means you can take off some winter layers, but there are tradeoffs. For one, the nights are growing shorter. In March, you can still stargaze as early as eight o'clock, but by May the sky isn't dark enough until ten. Also, Orion and his gang are making their celestial exits in the west. Earth, in its perpetual orbit around the sun, is gradually turning away from the brightest constellations of the year.

Don't get me wrong here. The spring heavens still contain many treasures, including the constellations Leo the Lion and Boötes the Farmer. Look for the bright star Arcturus in Boötes, a sure sign that you won't be wearing that winter coat much longer!

The Realm of the Galaxies (Photograph © Robert Gendler)

LEO THE LION

A giant lion roams the skies, leading the charge of spring constellations in the east. There isn't much question as to how to find Leo the Lion. In the early evening, just look for a large, backward question mark in the east. In the old days, the question mark was referred to as the Sickle. It lost that nickname when the sickle, once used for cutting grain or brush, was replaced on farms by sophisticated tractors and in the suburbs by electric weed whips.

Punctuating the bottom of the question mark is Regulus, the brightest star in Leo. Regulus is a slightly distant celestial neighbor, about 78 light years away. With a girth of over 4 million miles, Regulus is more than four times larger than the sun. This large, bright star marks the heart of the mighty lion, while the rest of the question mark outlines the head.

Look for a triangle to the lower left of the question mark—that's Leo's rear end and tail. The moderately bright star Denebola, 36 light years distant, marks the end of the beast's tail.

Celestial Goodies

Algieba: This star, the third-brightest in the constellation Leo, is a close binary system. One star shines with an orange hue, the other a yellow cast. The stars revolve around each other every 619 years. You'll need a moderate to large telescope to split it.
Magnitude: 2.0

M65 and M66: These faint twin spiral galaxies each measure about 50,000 light years in diameter, separated by 125,000 light years. You'll need a moderate to large telescope to see them.
Magnitude: 10 to 10.5

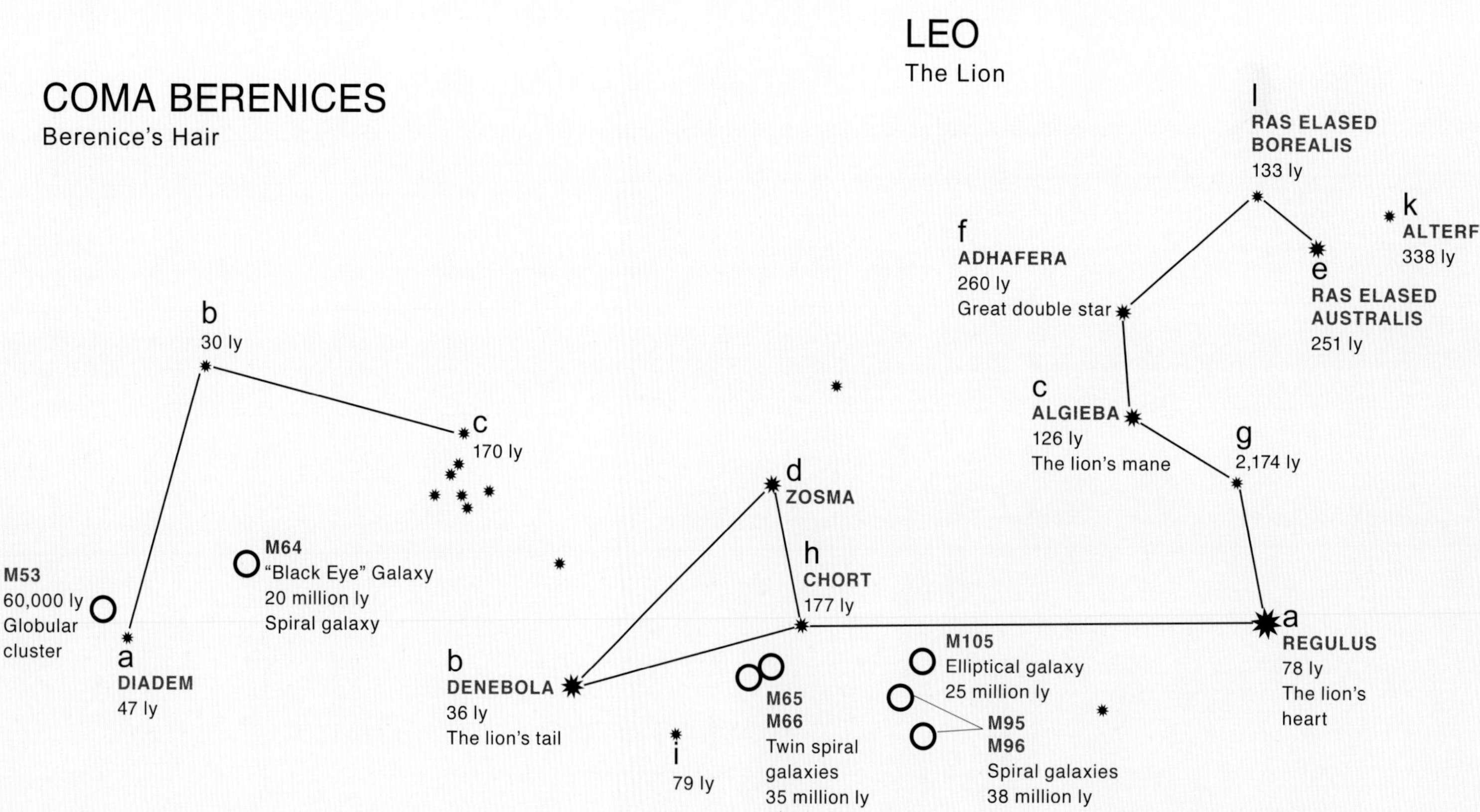

M95 and M96: This is another dim pair of spiral galaxies that you'll need a moderate telescope to see.
Magnitude: 10.5 to 11.0

M105: This elliptical galaxy, located near M95 and M96, is even fainter than the others in Leo.
Magnitude: 11.5

M44: The Beehive Cluster, otherwise known as the Praesepe, is a wonderful open cluster barely visible to the naked eye but seen nicely through binoculars or a low-magnification telescope. You see about fifty stars spread across an area over three moon-lengths wide. Most of the stars have a bluish tint, but some brighter ones have an orange hue. Look for the Beehive Cluster on the monthly star maps to the upper right of Leo's head.
Magnitude: 4.0

THE SUPER KING OF THE BEASTS

I love the Greek myth of Leo, the champion of ferocious lions in this part of the galaxy. To tell the tale of Leo, though, I have to introduce the famous Greek hero Hercules.

Hercules was equal parts muscle and heart. He sported a huge smile and found good in everyone. At an early age, he fell in love with the beautiful but conniving Princess Megara. They married in an ornate palace ceremony, and at first, things couldn't have been better. But soon, they began to argue. Their arguments turned into little fights that eventually turned into big fights. Princess Megara was a nitpicker. Nothing Hercules did ever pleased her highness and she picked and picked and picked until the gentle giant lost his sanity. In a fit of rage, he murdered Megara and all her attendants.

After the slayings, Hercules came to his senses and realized the horrible crimes he had committed. Full of remorse, he turned himself into the palace authorities. His fate was left to the mercy of Eurystheus, the king of Mycenae, who easily could have had Hercules beheaded for killing his daughter. But the wise king realized that his son-in-law had gone temporarily insane and he instead ordered Hercules to atone for his sins by performing twelve great labors.

His first labor was to kill Leo, the mightiest of lions, who for years had terrorized the kingdom. Leo devoured many of King Eurystheus's subjects. This lion was so tough that hunting spears bounced off his thick hide.

Hercules stalked the lion for weeks until finally an opportunity presented itself. After gobbling down a fair maiden for a mid-afternoon snack, the beast settled in for a nap. Hercules gave the lion a few minutes to doze off, then dive-bombed Leo from the rear and grabbed his thick neck. Using every ounce of his strength, Hercules struggled for hours, eventually choking the lion with his huge, bare hands!

Twin galaxies in Leo, M65 and M66 (Photograph © Robert Gendler)

The gods on Mount Olympus commended this heroic accomplishment. After Hercules completed his other eleven great labors and had paid his debt to society, the gods placed both Leo and Hercules in the night skies. Our great hero is fainter and much less detailed than Leo, and he dances mainly in the summer and early fall evening skies.

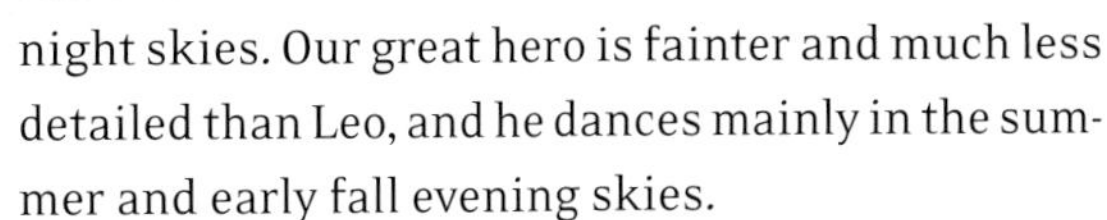

COMA BERENICES

Coma Berenices is one of those not-so-spectacular spring constellations. It's small and dim, but it's easy to find and has a great story. It's also one of the newer constellations, listed in 1602 as a constellation to honor Tycho Brahe, the infamous astronomer who had died just one year earlier. The son of wealthy Danish nobility, Tycho became fascinated with astronomy when he witnessed a partial solar eclipse at the age of fourteen. He marveled at the accuracy of the predictions for the time and the extent of the eclipse. He then used his mom and dad's money to set up a great astronomical observatory on a Danish island. When his money ran out and the king wouldn't subsidize his operation, he moved to

Prague at the invitation of the monarchy. His years of observations, without a telescope, went on to help many future astronomers, particularly Johannes Kepler, who used Tycho's observations to define his three laws of motion that apply to the planets revolving around the sun.

Tycho was not your run-of-the-mill ivory-tower type. He ate, drank, and partied hardy. He was also arrogant and combative. At age twenty, he became embroiled in a barroom sword fight and got his nose cut off! He replaced it with an artificial nose made of solid gold. Tycho and his gold nose met their demise at age fifty-five, when he died of a bladder infection after boozing it up at yet another party. He worked hard and played hard—a little too hard!

I got a little sidetracked here, but I love telling the story of Tycho Brahe. Anyway, the constellation named in his honor, Coma Berenices, means Berenice's Hair. It looks like faint strands of hair flowing near the zenith after twilight. You need to be in the countryside to really see it. Most of the spring, it's still high in the sky. It's not far from the Big Dipper's handle and the tail of Leo the Lion.

Black Eye Galaxy, M64 (Photograph © Jack Newton)

Celestial Goodies

M64: The faint Black Eye Galaxy shines with the luminosity of 10 billion suns. A dark patch makes it look, to some sharp-eyed observers, as if this galaxy has a black eye! You'll need a moderate to large telescope to see it. Magnitude: 9.0

M53: This globular cluster shows a lot of detail, despite its great distance from the earth. It's believed to be made of 100,000 stars crammed in a sphere about 300 light years wide. You'll need a moderate to large telescope to see it. Magnitude: 8.5

A HAIR-RAISING TALE

According to mythology, Coma Berenices is named after Queen Berenice, the wife of the famous Egyptian Pharaoh, Ptolemy III. The story goes that the great pharaoh was leading his troops in a fierce war, and every night Queen Berenice prayed to the gods for his safe return. She was so desperate to see him again that she promised to cut off all her beautiful hair if her husband returned safe.

A year later, Ptolemy returned victorious, and true to her word, the queen cut off her hair and dedicated it to the temple of Aphrodite, the goddess of love. Days later, souvenir-seeking scoundrels hoisted Berenice's hair out of the temple. When the heist was discovered, Ptolemy and Berenice were ready for heads to roll, literally! The temple priests were within hours of execution when a traveling group of Greek astronomy consultants literally saved their necks. They convinced Ptolemy and Berenice to go out with them that night to see a brand new pale cluster of light high in the sky. "Look!" one exclaimed, "Do you not see the clustered curls of the queen's hair?" Another continued, "Aphrodite and the other gods believed that the queen's hair was just too beautiful for a single temple to possess. Berenice's hair belongs in the heavens for all to see!"

To the relief of the temple priests, Berenice and Ptolemy swallowed this line of bull. Consultants can be very convincing, even back then!

BOÖTES THE FARMER

The constellation Boötes (pronounced boo-oat-tays) is supposed to be a hunting farmer. As with most constellations, it doesn't come close. Boötes actually looks like a giant kite rising on its side, in the east on a spring evening. It's easy to spot. Just remember to "arc to Arcturus." Look for the Big Dipper in the high northeast sky and extend the arc of the dipper's handle. You'll run into the bright star Arcturus. It's about 30 degrees, or three fists-width at arm's length, from the dipper's handle. You can't miss it. Arcturus is the brightest star in Boötes and the second-brightest star in our night skies.

Arcturus marks the tail of the sideways kite that's leaning to the left. Look to the left of Arcturus and see if you can spot the rest of the kite. By the way, your eyes are not deceiving you. Arcturus has an orange tinge. It's about thirty-four times the size of the sun, a small red giant star 37 light years away.

Another crazy thing about Arcturus is that it's orbiting our Milky Way at a right angle from the plane of our galaxy. Most of the stars in our galaxy revolve around the same plane of our Milky Way. Because of that, Arcturus is moving in Earth's direction at well over 2,000 miles a minute! Don't worry about a collision, though. It won't come anywhere near the earth and, in about half a million years, it will be so far away we won't see it anymore. Well, our great, great, great grandkids won't see it anyway!

Celestial Goodies

M3: This globular cluster, 200 light years in diameter, contains more than 500,000 stars. You need a moderate telescope to find and appreciate it. Magnitude: 7.0

BOÖTES
The Farmer

c SEGINIS 85 ly
b NEKKAR 219 ly
M3 35,000 ly Globular cluster
d 117 ly
g MUPHRID 37 ly
e IZAR 210 ly
a ARCTURUS 37 ly

h
b NUSAKAN 114 ly
i
a ALPHECCA 75 ly
c 145 ly
e 230 ly
d 166 ly

CORONA BOREALIS
The Northern Crown

THE GREAT HUNT

According to legend, Boötes was the son of Demeter, the goddess of agriculture. His father was a mortal man with whom Demeter fell in love. One thing led to another and, oops, Boötes was born to the unwed couple, making him a half-god. Back in those days when that sort of thing happened, these godly love-children were placed into adoption. These decisions were made more for convenience than shame. The gods were too busy running the heavens to raise their "accidental" children.

The goddess of agriculture placed Boötes, his half godliness, with a wealthy farm family, and for the first few years everything was great. Unaware that he was half divine, Boötes grew up alongside his mortal foster brother, hunting and fishing the surrounding woods and helping on the farm. The crops were good and the profits high. Then tragedy struck. Boötes's foster parents were killed in a chariot accident on New Year's Eve. They willed their money and the farm to Boötes and his brother, who, as the oldest son, served as executor of the will. Things would have been okay, except that the big brother was a crook. A month after the accident, he and his girlfriend took all the money and went on a global spending spree.

Boötes was broke and on his own. It was a real struggle. That spring on the farm, Boötes did all the tilling by himself, since he had no money for hired help. Back then, farmers had to work the land by hand. He kept thinking that there must be a better way to till the land. It was a combination of godliness, ingenuity, and desperation that led him to invent the plow that could be pulled by an ox.

Boötes really had something! Not only could he plow faster, but there was less wear on his body. Other farmers saw this new invention and wanted Boötes to build ox-pulled plows for them. Word spread farther, and pretty soon Boötes had a booming business. He sold the farm and concentrated on his plow business. He was loaded! As the business got better and better, he was able to take time for his true passions, hunting and fishing.

The gods on Mount Olympus learned of Boötes's accomplishments and eventually word reached Demeter, his birthmother. She was proud of her son and, as the goddess of agriculture, especially pleased with his contribution to farming. When Boötes was getting-on in years, Demeter gave her son the ultimate reward. She plucked him off the ground and placed his body in the stars as the constellation we see today.

Every summer and fall, Boötes has the ultimate hunting experience. He hunts the Big Bear, Ursa Major. It's said that Boötes is one of the happiest constellations in the sky. The hunt begins every spring with the Big Bear rising in the east and Boötes right behind him. By autumn, Boötes has nailed the bear with his bow and arrow. That's why the wounded Big Bear sinks close to the northern horizon with Boötes in hot pursuit. It's also said that the leaves turn red every fall from the bear's dripping blood. All right, the lack of chlorophyll may have something to do with it too.

Ursa Major is one tough bear, though, because he manages to lick his wounds and rise high in the northeastern skies every spring, with Boötes right on his tail. The great hunt goes on and on and on!

CORONA BOREALIS

Three challenges are posed when trying to spot Corona Borealis, otherwise known as the Northern Crown. First, this is a small constellation. Second, it's not bright, but you can see it with the naked eye if there isn't too much light pollution. Third, it doesn't look like a crown. Corona Borealis resembles a cereal bowl. Luckily, Corona Borealis is just to the upper left of a bigger and brighter constellation, Boötes the Farmer.

There isn't much to the Northern Crown. Many amateur astronomers I know refer to it as "Core Bore." The brightest star, located at the bottom of the bowl, is Alphecca (pronounced al-feck-ah). Alphecca is a hot bluish-white star, four times the size of the sun and about 75 light years away from Earth, which is right down the block, astronomically speaking.

THE LITTLE CROWN WITH THE BIG STORY

According to Greek mythology and blended legend, Corona Borealis is the crown of Princess Ariadne, the daughter of the evil King Minos of Crete. Minos was the Saddam Hussein of his time.

For sick kicks, King Minos sacrificed seven young men and seven young women once a year to his pet monster, the Minotaur. Minos's pet was a beauty. It had a bull's body and an incredibly ugly human head. On the same day every year, fourteen young people were led, one by one, to the Minotaur, and to their certain demise. Every year, Princess Ariadne was forced to watch this sick ritual. One year, Ariadne made eye contact with Theseus, one of the handsome young men being led to slaughter. He held up his chin and gave the princess a quick, strategic wink. That's all it took. Ariadne fell in love. She scrambled back to the castle, hid a sword under her cape, and raced back to the castle grounds, where she slipped the sword to Theseus when the guards were distracted by a group of local ladies.

When Theseus was led into the Minotaur's pen, out came the sword and he turned the monster into Swiss cheese. With the guards in shock, Theseus bolted to the castle gates where Ariadne was waiting. The couple stole a chariot parked in the market square and headed for the sea. There, they grabbed a boat and rowed their behinds off. As darkness fell, they stopped at the Island of Naxos for the night.

No one really knows what happened after that. Maybe Theseus got cold feet or maybe it was Ariadne's snoring, but whatever the reason, Theseus ditched the princess and left her sobbing on the beach of Naxos.

It just so happened that Bacchus, the god of wine, ruled Naxos. The god with the biggest wine cellar around took pity on Ariadne and convinced her that Theseus was a bum. He told her she could do better. This bachelor god thought of himself as the better. He had his servant fetch the finest wines out of his cellar, and after a few bottles, Bacchus and Ariadne were head over heels in love.

On their wedding night, Bacchus, that wine-sipping romantic, gave Ariadne a special gift. He took off his crown and, with all his might, threw it high in the night sky. It flew so high that it sprouted stars. Bacchus proclaimed that it was Ariadne's eternal crown, symbolizing their everlasting love. Hey, guys, top that!

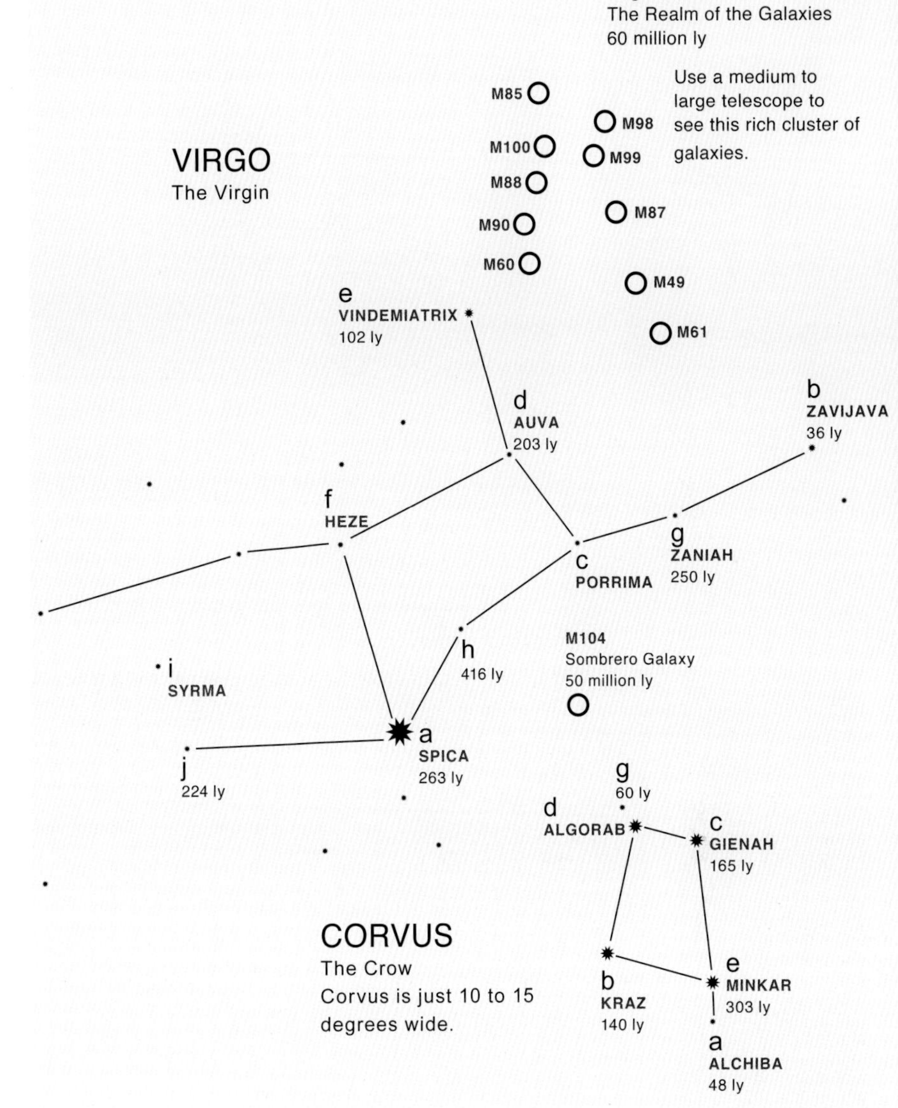

VIRGO THE VIRGIN

We see mainly animals, men, and gods in the night sky. Few women live in the celestial sphere. In fact, from the northern hemisphere, we see only two, Cassiopeia the Queen and Virgo the Virgin.

I know you've gazed at Cassiopeia before, although you may not have known what you were seeing. Cassiopeia is the bright "W" visible in the northern sky every night as it makes a tight circle around Polaris every twenty-four hours.

The other lady of the night skies is not so prominent. Virgo is a large but faint constellation, the kind I call a stargazing "deep track." It is a mystery to me how this faint collection of stars, about halfway up the southwestern sky, is supposed to be the goddess of fertility. It has only one bright star, Spica, and the rest are so faint that there is no way you can see them in the lights of the city, or even the outer suburbs. You need to be out in the countryside, armed with patience.

The best way to find Virgo is to find Spica, and that's easy. Look in the high northwestern sky for the Big Dipper, hanging by its handle. Follow the arc of the handle and you'll run into the bright-orange star Arcturus, the brightest star in the constellation Boötes the Farmer. Continue the arc beyond Arcturus and the next brightest star you'll happen upon is Spica, which marks the left hand of Virgo.

Spica is a blue giant star about 263 light years away from the earth. It's ten times more massive and over ten times larger than the sun, with a girth of almost 8.5 million miles. Spica is also hotter, with a surface temperature over 36,000°F. Spica is actually a close binary system. One star is believed to be a dying giant star, while its companion is about the size of the sun. They are practically touching, separated

by less than 12 million miles, and spinning around each other every four days. I'm surprised they don't get dizzy.

If you get a chance to go to the countryside on a clear, moonless night, scan the area between the constellations Virgo and Leo with a telescope and you're bound to see at least a few fuzzy stars and patches. You're seeing just a few of the brighter galaxies that make up the famous Virgo Cluster of galaxies, called the Realm of the Galaxies. There are over a thousand galaxies in the Virgo Cluster, with billions and billions of stars in each galaxy. Just think, those little fuzzies you're seeing are over 60 million light years away!

Celestial Goodies

M49 : This elliptical-shaped galaxy is one of the brighter of the Virgo Cluster. Use a moderate to large telescope to view it.
Magnitude: 10.0

M87: This huge, elliptical-shaped galaxy is one of the brightest in the group. It's estimated to be over 120 light years across and packed with several trillion stars. Wow! You'll need a small to moderate telescope to see it.
Magnitude: 8.1

M104: The Sombrero Galaxy in the Virgo Cluster faces Earth edge-on, giving it the look of the famous Mexican headdress. You'll need a moderate to large telescope to see it.
Magnitude: 9.5

BELOW: Sombrero Galaxy, M104 (Photograph © Russell Croman)

THE LONG-SUFFERING WIDOW

To many cultures, including the Greeks and Egyptians, Virgo the Virgin represents the goddess of fertility. She holds in her hand a shaft of wheat. In fact, farmers took the first sighting of Virgo, with its bright star Spica, as a cue to start spring planting. When she leaves the evening sky four to five months later, the growing season is over. According to mythology, that's when Virgo leaves the land of the living and begins her annual search in the underworld for her slain husband, Tammuz. She hasn't found him yet, but every year she resumes her search. The grand lady of the night sky is a loyal lover!

CORVUS THE CROW

The constellation Corvus the Crow may be small and insignificant, but it's still one of my favorite little constellations. Look for it in the spring, just after twilight around nine o'clock, in the low southeastern sky. Will you see a soaring crow lit up like a neon sign? No, but you will see a distinct little lopsided diamond of moderately bright stars. If you have trouble finding it, first look for the brightest star in that part of the sky, Spica. Corvus is just to the left.

THE CROW WHO ATE CROW

According to Greek mythology, Corvus brought disgrace to all crows after he failed to carry out a mission. Back then, crows were respected birds that served as messengers, or "go-fers," for the gods. They were beautiful, sociable birds with white feathers and gold-trimmed wings. They also sang more beautifully than any other bird. The sociability part is what got Corvus in trouble.

Apollo, the god of the sun, sent Corvus on a mission to fetch some water from a far-off magical fountain. Apollo gave Corvus a special cup to carry the water. It was a hot, dry Friday afternoon when Corvus flew off in search of the fountain. He flew for hours and hours, got lost, and, like most male crows, didn't stop to ask for directions. He just flew and flew, toting that cup.

Around early evening, his wings grew tired and his throat dry. In the distance, Corvus saw a bar and decided to take a break. That's when the real trouble started. In the bar, Corvus ran into some buddies he hadn't seen in years. You know what happened next. One beer turned into two beers, which turned into three, and so on. Even worse, Corvus chugged his beer out of the cup Apollo had sent along to fetch the water. After a night drinking and laughing, Corvus was in no shape to fly and had to sleep it off in a local park.

The next morning, a hung-over Corvus resumed his search for the magical fountain. Once again, he flew and flew but to no avail. He just couldn't find his destination. He decided to fly home and face Apollo. All the way, he dreamt up different excuses for his failed mission. He finally concocted a story about being attacked and bit by a crazed water snake, leaving him too woozy to find the fountain.

As Corvus approached the mountain, he saw Apollo waiting for him. The god of the sun scowled when the crow landed with an empty cup. Corvus tried his story about the snake and had Apollo just about convinced, until he handed back the cup. Apollo took a close look to see if it was damaged. That's when he smelled the stale beer.

Apollo was so mad he fired Corvus on the spot and got together with the other gods to have crows banished from Mount Olympus. It didn't stop there. Crows lost their gold-trimmed white feathers and were turned into jet-black birds. As a final punishment, crows lost their melodic singing voices, which were replaced by the cawing we hear today.

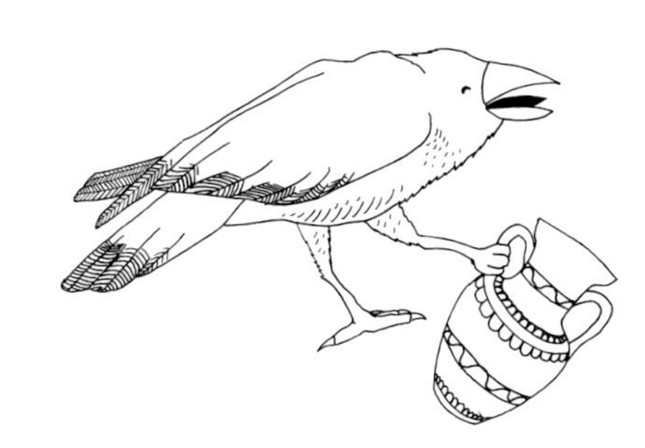

10

SUMMER CONSTELLATIONS

Summer is a great time to take a road trip to the country for some serious stargazing. You need the dark air to get the most out of the summer sky. Bring along lawn chairs, snacks, and bug juice. Better yet, make a camping trip out of it! You'll need that afternoon nap, though, because the sky isn't dark enough to make the stars your old friends until after ten o'clock.

In the country, you can't help but see the Milky Way, the white band running north to south in the summer sky. All the visible stars in our sky are part of the Milky Way Galaxy, but the ghostly band known as the Milky Way consists of the combined light of the distant stars in the plane of our galaxy. The heart of the Milky Way, about 70,000 light years distant, is visible in the summer constellation Sagittarius.

Also look for the Summer Triangle. It contains the brightest stars in the summer sky: Vega in the constellation Lyra the Lyre, Altair in the constellation Aquila the Eagle, and Deneb in the constellation Cygnus the Swan. Just look for the three brightest stars in the eastern heavens, and that's the Summer Triangle!

The Milky Way running through the constellation Cygnus (Photograph © Thomas Matheson)

HERCULES THE HERO

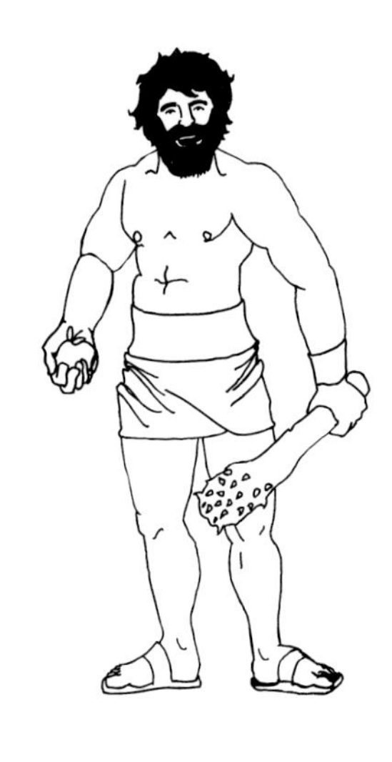

The first of the summer constellations is not the brightest; in fact, it's one of the dimmer constellations seen throughout the year. Hercules starts in the high eastern sky at the beginning of summer and, by the end of August, has migrated to the high western sky. It is supposed to represent the figure of a hero hanging upside down, but it looks more like a fancy handwritten "H," for Hercules. A large trapezoid of stars, visible to the naked eye in reasonably dark skies, makes up the center of the constellation.

The best part of Hercules is a star cluster located on one side of the center trapezoid. It's the brightest globular cluster in our skies. Half a million stars are jammed into a sphere held together by mutual gravity, 140 light years in diameter and more than 25,000 light years away. Most astronomers believe it's one of the oldest objects in the sky, possibly over 13 billion years old. Many think it formed shortly after the Big Bang, the theoretical start to the universe.

Celestial Goodies

M13: The Hercules Cluster is the best and brightest globular cluster in our sky. It's a "must see" with a small to moderate telescope!
Magnitude: 7.0

M92: This is another wonderful globular cluster, but it doesn't generate much respect next to the glorious M13. Made up of 300,000 stars, this cluster is over 13 billion years old. It's worth a look-see with a small to moderate telescope.
Magnitude: 7.0

Rasalgethi: View this star, the brightest in Hercules, through a small to moderate telescope and you'll see a close double star. One is orange, the other is bluish-green.
Magnitude: 2.8

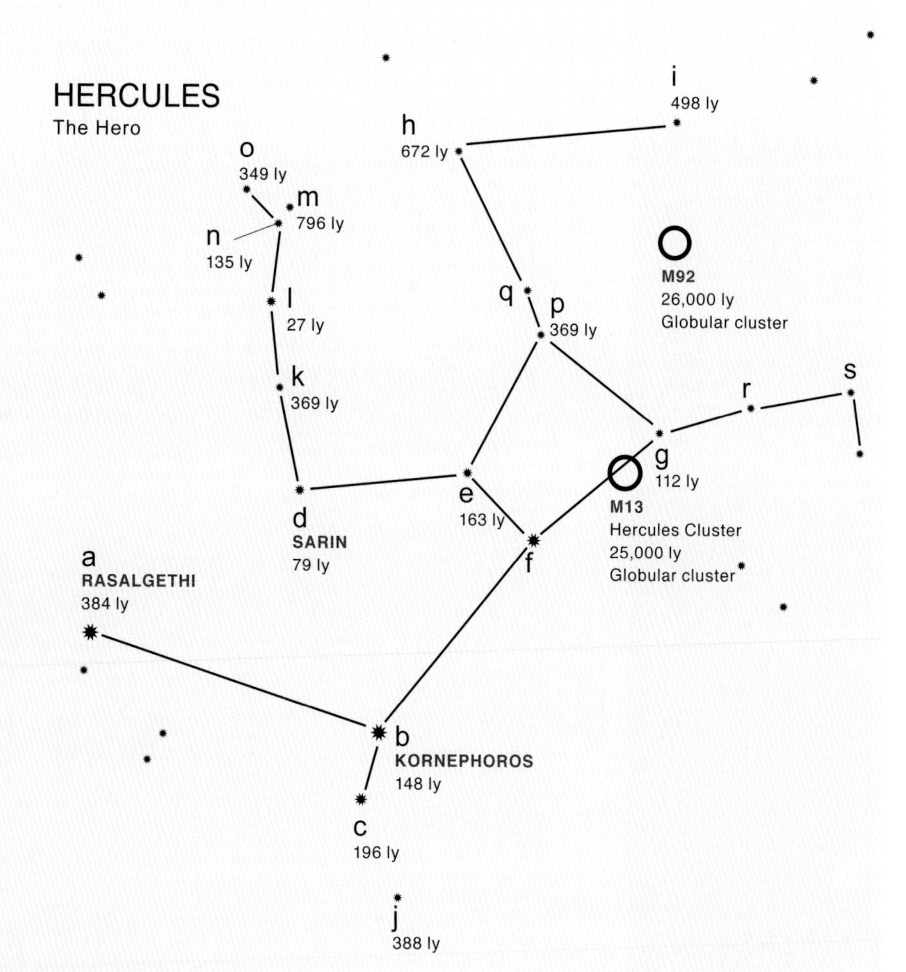

THE REPENTANT HERO

The legend of Hercules the Hero is covered in the Greek mythological story of Leo the Lion. Hercules is best known for performing twelve great labors, a penance for killing his wife in a fit of rage. The first labor was to slay Leo, the king of the beasts. Using his strength and his brain, Hercules destroyed the monstrous lion. Over the years, he went on to complete all twelve of the assigned labors. Zeus and the rest of the gods rewarded Hercules at his death by placing his body in the heavens. They didn't want him to receive full honors because of his murder conviction, so he hangs upside down.

CYGNUS THE SWAN

Throughout the year, our night sky contains all kinds of critters, including eight birds. The biggest and brightest feathered constellation seen around here is Cygnus the Swan, flying in the eastern sky. The bright star that marks the tail of the swan is Deneb, one of the stars of the Summer Triangle. You'll find it in the mid to low eastern sky in early summer and, by summer's end, in the high eastern sky. The other Summer Triangle stars are Vega and Altair, the brightest stars in their respective constellations, Lyra the Lyre and Aquila the Eagle. Just look for the three brightest stars in the eastern heavens and you've found the Summer Triangle!

Deneb marks the left corner of the triangle. It's the dimmest star of the three, but hardly puny. Quite to the contrary, Deneb is a fantastically huge star. Some astronomers think it's over 3,200 light years away. It's so far away that the light we see from it tonight left that star around 1200 B.C. Theoretically, it could explode tonight and our great, great, great, great . . . grandkids would see the explosion in A.D. 5000!

The fact that we can see Deneb with the naked eye says that it must be one enormous star. According to the latest data, Deneb is over 220 million miles in diameter and over 300,000 times the luminosity of the sun. Our sun is only 865 ,000 miles in girth. If we could magically pull Deneb 100 times closer to Earth—to a distance of around 32 light years, the approximate distance of the star Vega—Deneb would shine nearly as bright as the full moon. We would easily see its light in the daytime.

The constellation Cygnus contains an asterism called the Northern Cross. In fact, it's easier to see the cross than the swan. Deneb marks the head of the cross, which leans to the left. At the foot of the cross is the not-so-impressive star Albireo—not impressive, that is, to the naked eye. With a small telescope, you will see Albireo is a beautiful pair of stars, one gold and the other blue. It's one of the best double stars in the sky.

To expand from the Northern Cross and find the entire swan is easy: just look for the stars off either end of the arms of the cross and turn them into wings. Deneb becomes the tail of the giant swan and Albireo the swan's head.

Celestial Goodies

Albireo: This star is the best double in the night sky. One is orange and the other is blue. The stars orbit each other over a period of about 100,000 years. Don't wait up!
Magnitude: 5.1

Hercules Cluster, M13 (Photograph © Thomas Matheson)

NGC7000: The good news about the North America Nebula is that it's best seen with the naked eye, but the bad news is that you need super-dark skies. Long-exposure photographs show that it roughly takes on the shape of North America.
Magnitude: 4.0

M39: This nice open cluster contains thirty to forty stars. You might spot it with the naked eye, although you'll see it better with binoculars or a small telescope.
Magnitude: 5.5

NGC6960: The Veil Nebula is the faint remnant of a supernova explosion that took place thousands of years ago. You need a really dark sky and a large telescope to see it. Good luck!
Magnitude: 12.2

NGC6826: This faint planetary nebula contains a bright central star called the Blinking Planetary because its brightness can overwhelm your eyes so much that you won't see the surrounding nebula. When you use averted vision, the nebula blinks back. Large telescopes show a greenish tinge.
Magnitude: 9.8

M27: The Dumbbell Nebula is the brightest planetary nebula in the sky. With binoculars, it looks like a fuzzy patch, but a decent telescope reveals its hourglass or dumbbell shape.

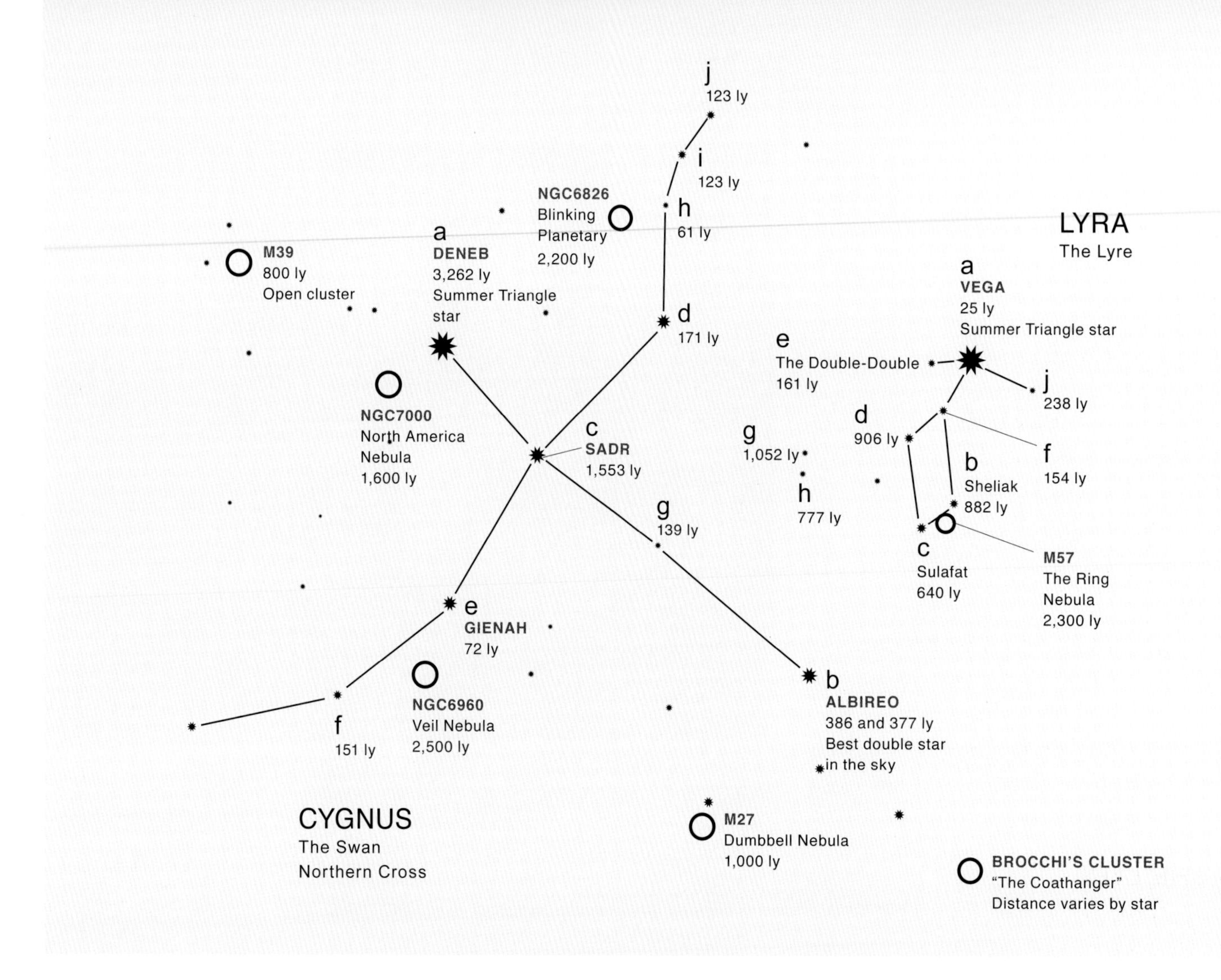

You might have a hard time seeing the central star, which is turning into a white dwarf, but you will surely see what it's blowing off! The nebula is thought to be over 2 light years in diameter. Magnitude: 7.5

Brocchi's Cluster: This cluster is nicknamed "The Coathanger," because through binoculars or a small telescope it looks exactly like a coathanger. Even though the stars line up neatly, they are at various distances from the earth and have nothing to do with each other. It's a random arrangement. Go see it. You'll like it! Magnitude: 5.1

THE GREAT DIVE

The story of how Cygnus got in the sky is a sad one involving Apollo. According to Greek mythology, Apollo was one of the most important gods on Mount Olympus. After all, he was the god of the sun, with the important job of guiding the sun chariot across the sky. The glass chariot was pulled gallantly by a fleet of flying white horses. Apollo loved his job and was rewarded handsomely by Zeus, the king of the gods.

In his youth, Apollo had been quite a playboy, but he eventually settled down, for the most part. He married and had lots of kids—in and out of marriage. Let's put it this way: Apollo was no saint! One of his kids was Phaethon, who at ten years old idolized his dad and wanted to take over the reins when Apollo retired. Phaethon begged his dad to let him take the sun chariot for a ride, but Apollo said no. Phaethon was just too young. Phaethon, though, was convinced he could handle it. He often rode along with his father and studied his driving techniques. He just knew he could do it.

One morning, Apollo's alarm clock didn't go off. Phaethon was up early that morning and saw that the sun had failed to rise. At first, he rushed to roust his dad out of bed, but then he saw the sun chariot inside its hanger. Temptation set in. This was his chance! He climbed in the chariot, backed out of the hanger, and bellowed "giddy up" to the flying horses. Before he knew it, he was airborne.

Phaethon would have been fine, but he started to hotdog it with the sun, zigzagging and pulling celestial wheelies. He soon lost control. From Mount Olympus, Zeus saw the careening chariot and thought some scoundrel had stolen it. He shouted to Apollo and woke him. While Apollo raced to borrow the moon chariot from his twin sister Artemis, Zeus shot a lightning bolt at Phaethon, spearing him out of the driver's seat. Apollo, unaware that his son was falling to his death, caught up with his sun chariot and soon had it under control.

Meanwhile, Phaethon swan dived into the river Po and drowned. Other gods recognized the body when it surfaced and took pity on him. They transformed his body into the beautiful constellation we see today, Cygnus the Swan. Phaethon is flying again!

LYRA THE LYRE

In the entire night sky, only one constellation is a musical instrument–Lyra the Lyre. In case you're not familiar with ancient musical stringed instruments, a lyre is the great, great, great grandfather of today's harp.

Lyra the Lyre is mostly a small, faint constellation, but finding it is a cinch because it contains Vega, the second-brightest star in the summer sky, at around 150 trillion miles distant. Look for the brightest star you can see in the high eastern sky–that's Vega! The only brighter star in the sky is Arcturus, an orange hued star in the high southwest. Launched in 1983, the Infrared Astronomical Satellite, which uses infrared radiation to see beyond obscuring gas clouds, detected a disk of dust and gas around Vega that may be a developing planetary system. Look for four faint stars that make up a parallelogram just to the lower right of Vega. Those four little shiners and Vega outline the ancient harp. Your guess is as good as mine as to how that's supposed to be a harp, but what intrigues me is that those four stars have nothing to do with each other. The stars vary in distance from 154 light years to over 900 light years away. They just happen to fall in the right line of sight to make that perfect parallelogram.

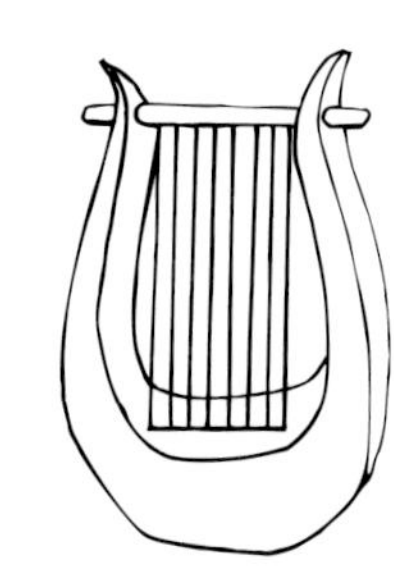

Celestial Goodies

M57: The famous Ring Nebula, which through a small to moderate telescope looks like a blue-tinted smoke ring, is the best planetary nebula in the night sky. It's a halo of gas being blown off a dying star. Some astronomers think this gas emerges from the parent star at 12 miles per second.
Magnitude: 9.5

Epsilon Lyrae: The Double-Double appears to be a single dim star to the naked eye, but it is actually a pair of double stars, four stars in all, separated by about 1.2 trillion miles. The north pair is separated by 1.3 billion miles and orbits each other every 1,000 years. The south pair is separated by just over a billion miles and orbits each other every 600 years. To add to this dance, the two pairs orbit each other every half-million years. You can see the Double-Double with a small to moderate telescope.
Magnitude: 4.0 to 5.0

THE GRIEVING SUPERSTAR

The constellation Lyra has many mythological stories. The version I like is the Greek and Roman tale about Mercury, the messenger of the gods, who

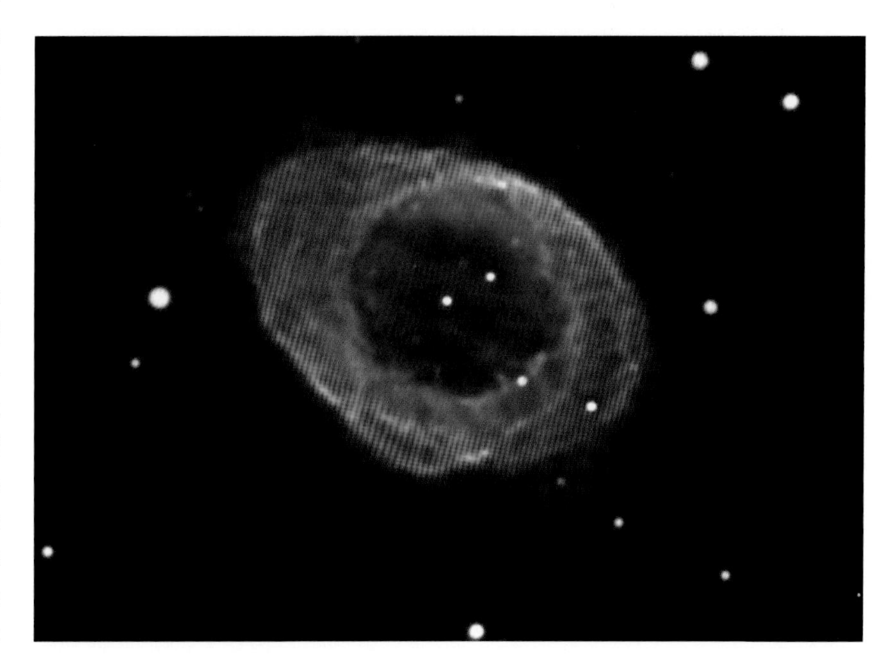

Ring Nebula, M57 (Photograph © Robert Gendler)

invented the first lyre from a tortoise shell and strings made of cow gut. Mercury was proud of his new instrument, but didn't have the talent to make beautiful music. He decided to give the lyre to Apollo, the god of the sun, as a birthday present.

At first, Apollo was thrilled with the lyre and practiced it every day. After awhile, the thrill wore off and he played it less and less, until it became a great dust collector in the sun god's palace. One day, Orpheus, Apollo's young son, picked up the harp, blew off the dust, and immediately made beautiful music with it. Orpheus was a natural! His music was so wonderful that wild animals came to listen and treetops bent over to hear him. Even fire-breathing dragons were lulled to sleep by the soothing tones.

Orpheus grew to be a handsome, talented man. He married the beautiful princess Eurydice and had a great life. He and his orchestra went on tour and commanded huge money for their concerts. Orpheus was the Bono of his time. He had palaces all over the countryside and the best chariots. Unfortunately, tragedy struck when Eurydice, his beloved, was fatally bitten by a poisonous snake.

The grief-stricken Orpheus went into seclusion for over a year, but finally pulled himself together enough to pick up his harp and resume his beautiful music. When he went back on tour, he was mobbed by fans. Now that he was single, young women threw themselves at him. He was in no mood to meet them. No one could replace Eurydice, the love of his life.

After one concert, he was sneaking out the back door when a mob attacked. As usual, he refused the women's advances. Security wasn't what it should have been, and the mob grew violent. The scorned women literally tore off his head and threw his body and lyre into a nearby river. Talk about a tough crowd!

Apollo and the rest of the gods recovered what was left of Orpheus from the river and buried him at the foot of Mount Olympus. They placed his magical lyre in the stars as the constellation we see today. It's said that if you are stargazing in the countryside, not only can you see Lyra the Lyre, but if you're quiet, you may hear its celestial tunes.

AQUILA THE EAGLE

Perched in the summer sky are a good bird and a bad bird. The good bird is Cygnus the Swan and the bad bird is Aquila the Eagle. This evil eagle soars in the southeast after evening twilight.

This constellation is easy to find because of its brightest star, Altair. Marking the cold heart of the winged beast, Altair is one of the three stars that make up the Summer Triangle. It is located on the lower-right corner of the triangle. The closest of the Summer Triangle stars to Earth, Altair is about 17 light years away.

Altair is not an average star. While the sun takes almost a month to complete one spin on its axis, Altair takes only 6.5 hours. There's no way to see this, even with the Hubble Telescope, but from spectral studies, astronomers figure that Altair's diameter at its equator is twice its diameter from pole to pole. Because of Altair's rapid whirl, the star is twice as wide as it is tall!

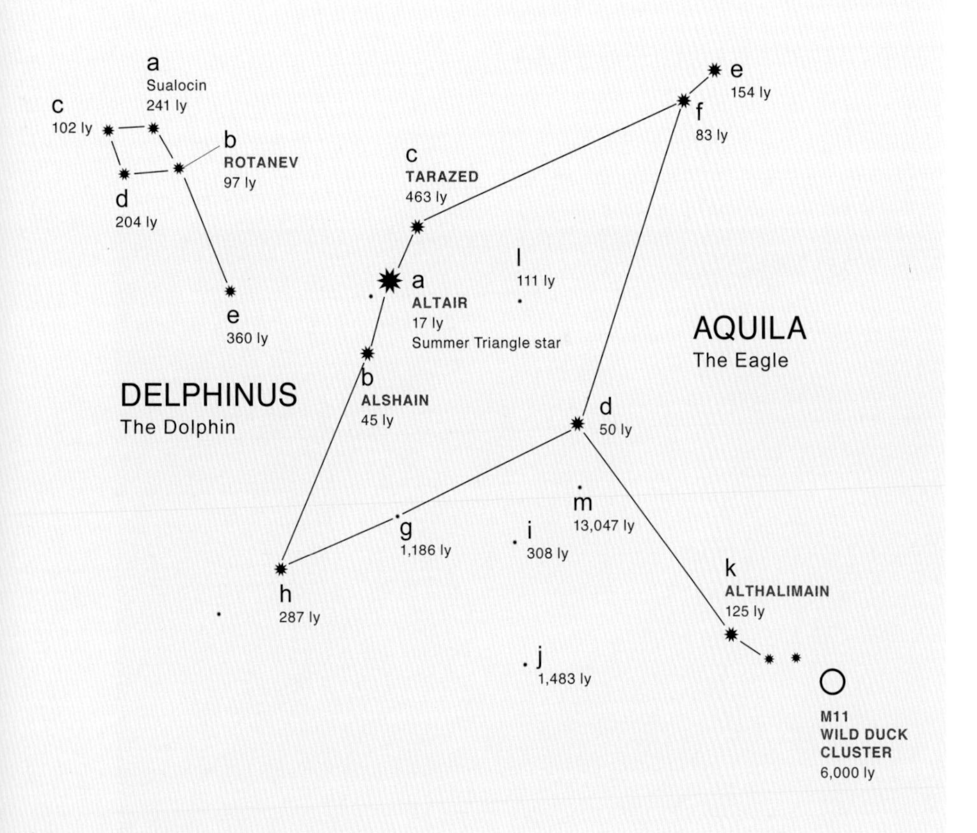

A giant vertical diamond of stars depicts the eagle's wingspan. A small line of stars, trailing off the right side of the diamond, depicts its tail. As far as this bad bird's head, well, that's left up to your imagination. It's supposed to be to the left of Altair, but nothing is there except some really faint stars. Maybe his head is invisible, like a Stealth Bomber!

Celestial Goodies

M11: The Wild Duck Cluster, a great open cluster of over 1,500 stars, is easily found with a small to moderate telescope. With your imagination, you can see a duck taking flight.
Magnitude: 7.0

BEWARE THE EVIL EAGLE

Aquila was Zeus's favorite pet. The king of the gods first noticed Aquila one day when the high-flying eagle soared over the palaces on Mount Olympus. The bird was not intimidated by the headquarters of the gods and he roosted on the gates. Zeus set out raw meat for his winged friend, and soon Aquila became his faithful pet. The bird served as a messenger and he even distributed thunderbolts across the sky.

Over the years, Aquila's duties widened as he became Zeus's "hit" bird. If Zeus had a bone to pick, his favorite eagle would do just that. Aquila had a razor-sharp bill and would literally pick the flesh off the bones of whatever or whomever he attacked. He'd swoop down and attack mortal and god alike with great dispatch and no mercy!

One of Aquila's victims was Prometheus, an old god from the Titan family. Prometheus had given humans the gift of fire some years back, which enraged Zeus, who felt fire was too great a gift to bestow on humans. As punishment, Zeus had Prometheus stripped naked, chained to a pillar, and every day around noon, Aquila would shoot out of the sky, peck through Prometheus's belly, and tear out his liver. Since Prometheus was immortal, his liver grew back overnight. The next day, Aquila would be back for another liver lunch. This torture went on for years!

The moral of the story: When you're stargazing on a summer evening, don't play with fire. Unlike Prometheus, your liver won't grow back quite as fast!

DELPHINUS THE DOLPHIN

Delphinus (pronounced del-fine-nus) is a tiny constellation, but this little dolphin is one of the gems of the night sky any time of the year. This constellation is also one of the few constellations that looks like what it's supposed to be.

As soon as the sky is dark enough, face east-southeast and look for the Summer Triangle. Delphinus is near Altair, the star that marks the triangle's lower right-hand corner. To the lower left of Altair, less than two fists-width at arm's length, try to spot a faint little sideways diamond of stars. This diamond outlines the dolphin's torso, and another star to the lower right of the diamond marks the tail.

One look at this constellation and you're likely to say "awesome" or "cute" or, as I hear from kids at my stargazing parties, "sweet." The constellation really looks like a leaping dolphin. Delphinus always reminds me of the ancient 1960s TV show *Flipper*, as well as the dolphin insignia on the helmets of the NFL's Miami Dolphins football team. With a telescope or a good pair of binoculars, you might see that the dolphin's nose is actually a double star, made up of two stars with a distinct yellowish tinge.

DOLPHINS TO THE RESCUE

The lore of how Delphinus got in the sky is as delightful as the little dolphin itself. One of the stories centers around Arion, a musical superstar of his time. He sang and played his harp all over the world, at least the world as it was then known. Everywhere Arion went, his fans tried to touch him, get his autograph, and just breathe the same air he was breathing. Street vendors made a bundle selling Arion togas and souvenir harps.

If airplanes had existed back then, Arion would have owned a 747, but since they were a few years distant, he traveled by land and sea. He had his own yacht and crew that rowed him around the Greek isles and beyond. Unfortunately, Arion was a

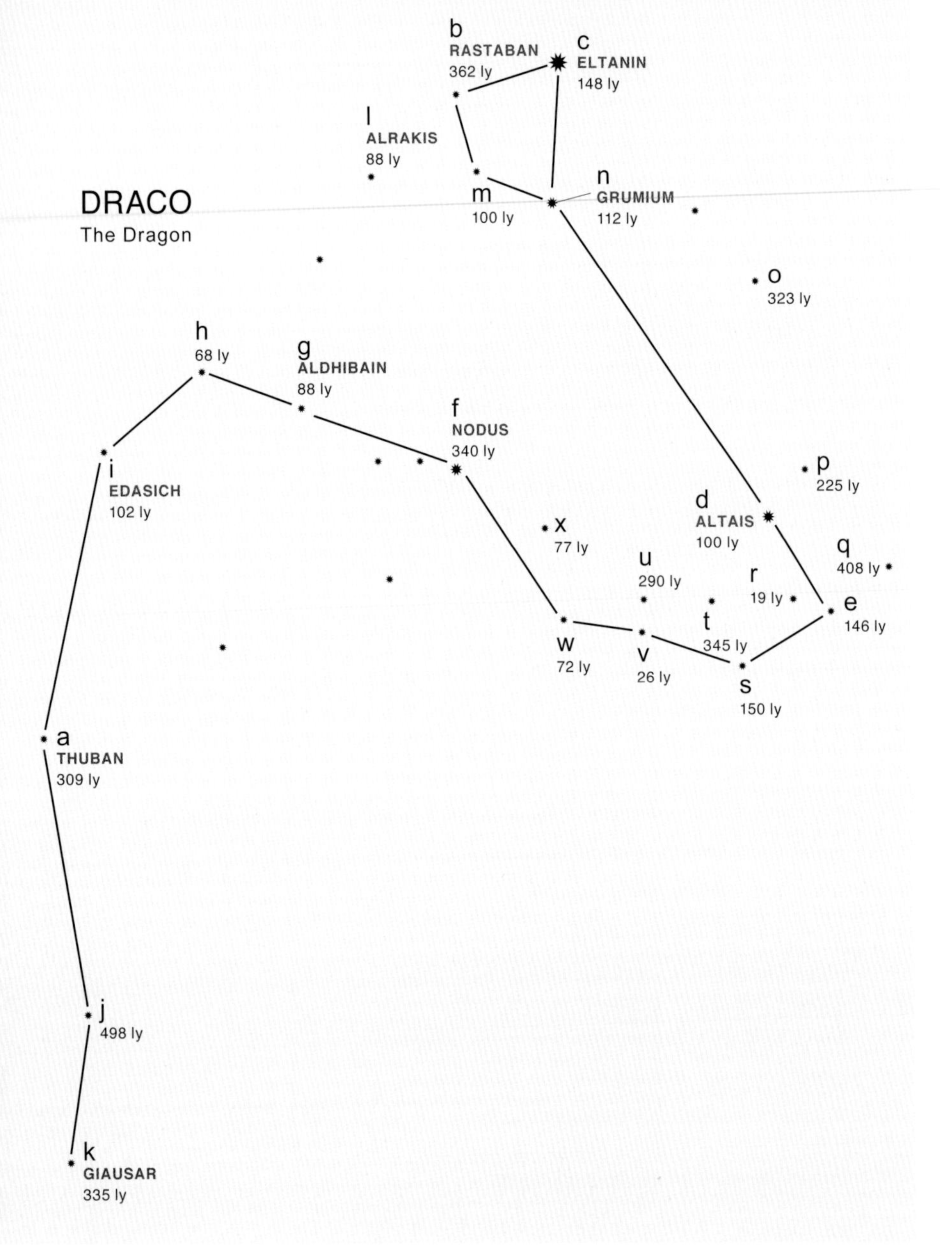

cheapskate when it came to paying his crew. Every night, the crew watched with growing envy as Arion, who demanded to be paid in cash, climbed aboard the ship with sacks of money.

Mutiny was on the minds of the crew, and one night after a concert in Sicily, it happened. They jumped Arion and put him on the plank for a final stroll. Terrified, Arion promised to give them all 50 percent raises, but it was too late. He was to walk the plank to his doom. Arion begged to play his magical harp just one more time. They threw his harp to him and yelled, "Okay, tightwad, sing before you swim." Arion clutched his harp and sang with all his might. His music was so beautiful that dolphins gathered below and sang along. Arion extended his song as long as he could, but when the final chorus ended, into the ocean he went.

Apparently, the dolphins hadn't heard enough and went to Arion's rescue. Delphinus, the largest of the group, hoisted the drowning Arion out of the water and onto his back. Delphinus gave Arion a ride all the way back to Greece, where Arion and his music lived on. The gods on Mount Olympus were so pleased to hear about Arion's rescue that they raised the great dolphin to the celestial sea in the sky, where he swims among the stars every night, still humming his favorite Arion tunes. If you listen on a quiet night, you might hear the singing dolphin!

DRACO THE DRAGON

Warm summer evenings are made for unwinding under the stars. Speaking of unwinding, one of the great summertime constellations is Draco. It's actually circumpolar in our sky, but it's best seen in summer and early autumn. Draco is supposed to be a dragon, but forget that! The winding beast looks more like a snake. It's not the brightest constellation, but it's a fun challenge to pick it out in the heavens. You'll feel a sense of accomplishment when you see its backward "S" shape. You don't need a pitch-dark sky, but it helps to be away from heavy city lighting.

The brightest part of Draco is the head, a distinct trapezoid of four stars high in the northern sky, near the overhead zenith. It's to the lower left of Vega, the brightest star in Lyra the Lyre. After you've spotted the head, look for the next two brightest stars to the lower right. From there, you'll see a line of stars that kinks to the upper left and then kinks again to the lower left. The tail of the dragon tapers off between the Big and Little Dippers.

The brightest star in Draco is Thuban, the third star from the end of the tail. Because of the slow wobbling of the earth's axis, Thuban used to be our North Star about 4,000 years ago. In fact, the great Egyptian pyramids at Giza may have been built with Thuban as a guide. One of the pyramids was constructed in such a way that you could see Thuban every night from the bottom of a deep airshaft. Egyptians took their stars seriously. These days, Polaris is our North Star, but the outstretched dragon is still close by.

THE WATCH DRAGON

According to legend, Draco the Dragon provided security for Hera, the queen of the gods. Loyal and vigilant, he guarded the sacred golden apples that Zeus gave his new bride at their wedding. Hera kept the golden apples in the exquisite Garden of Hesperdes. The faithful dragon guarded the apples day and night, not even taking time off to sleep.

Late one night, the Greek hero Hercules crashed the gate of the Garden of Hesperdes on a direct path to snatch the apples. It was one of his twelve great labors. Hercules actually had his hand on an apple when Draco whacked him across the face with his gigantic tail. A battle broke out that went on for hours. Draco just about had Hercules wrapped up, but the great hero managed to grasp a dagger hidden in his shoe and thrust it into Draco's heart. With the watch-dragon out of the way, Hercules loaded up his sack with golden apples and was off.

Hera was saddened by the loss of her apples, but nothing compared to her grief when she discovered the bloodied body of Draco. She honored her faithful dragon by flinging him into the northern skies. There was just one problem. Because of his wounds, when Hera tossed Draco into the heavens, his body stretched out, giving him the look of the snake that we see every night.

SCORPIUS THE SCORPION

When I see the constellation Scorpius the Scorpion after twilight, I know summer has kicked in. Scorpius is a rarity. It's one of the few constellations that almost looks likes what it's supposed to be. It also has a nickname: Fishhook. I remember my grandma pointing out the big fishhook from the dock of her cabin near Garrison, Minnesota.

Finding Scorpius does not require great astronomical effort. Look in the low southern skies for Antares, a reddish star that marks the heart of the scorpion. It's the brightest star in that part of the heavens. To the right of Antares you'll see three dimmer stars in a vertical row that make up the scorpion's head. To the lower left of Antares, look for the beast's long, curved tail. The scorpion's tail is easier to see in the southern United States, where it's higher in the sky, but you can see the tail from our area if you have a low tree-line and a flat southern horizon.

The name Antares alludes to the star's reddish hue. Derived from the Greek language, Antares means "rival of Mars," since the star has the same ruddy tone as the red planet. In fact, you can easily confuse Mars and Antares if you're new to stargazing.

There's no confusing Mars and Antares, however, when it comes to size. Mars is only about 4,000 miles across, a far celestial cry from the over 2-billion-mile diameter of Antares! That's over 2,500 times the diameter of our sun!

Celestial Goodies

M4: The Cat's Eye Cluster is one of the oldest globular clusters in our sky, at the ripe old age of roughly 10 billion years. It's about 100 light years in diameter and visible through a small to moderate telescope. It might be difficult to find, however, because light from the nearby bright star Antares can drown out the cat's eye. It also tough to see well because most of the time it's low in the southern sky, forcing us to look through more of our atmosphere.
Magnitude: 7.5

M6: The Butterfly Cluster is a decent open cluster of young stars about 100 million years old. The cluster roughly outlines the shape of butterfly. You can see it with a small to moderate telescope.
Magnitude: 4.5

M7: The Ptolemy Cluster is another nice open cluster almost visible with the naked eye. Its stars, numbering around eighty, spread over 20 light years.
Magnitude: 3.5

M80: This small globular cluster, only 50 light years across, is crammed with hundreds of thousands of stars. So much for privacy! The atmosphere blurs this low-lying cluster in our sky, so use a moderate to large telescope for best viewing.
Magnitude: 8.5

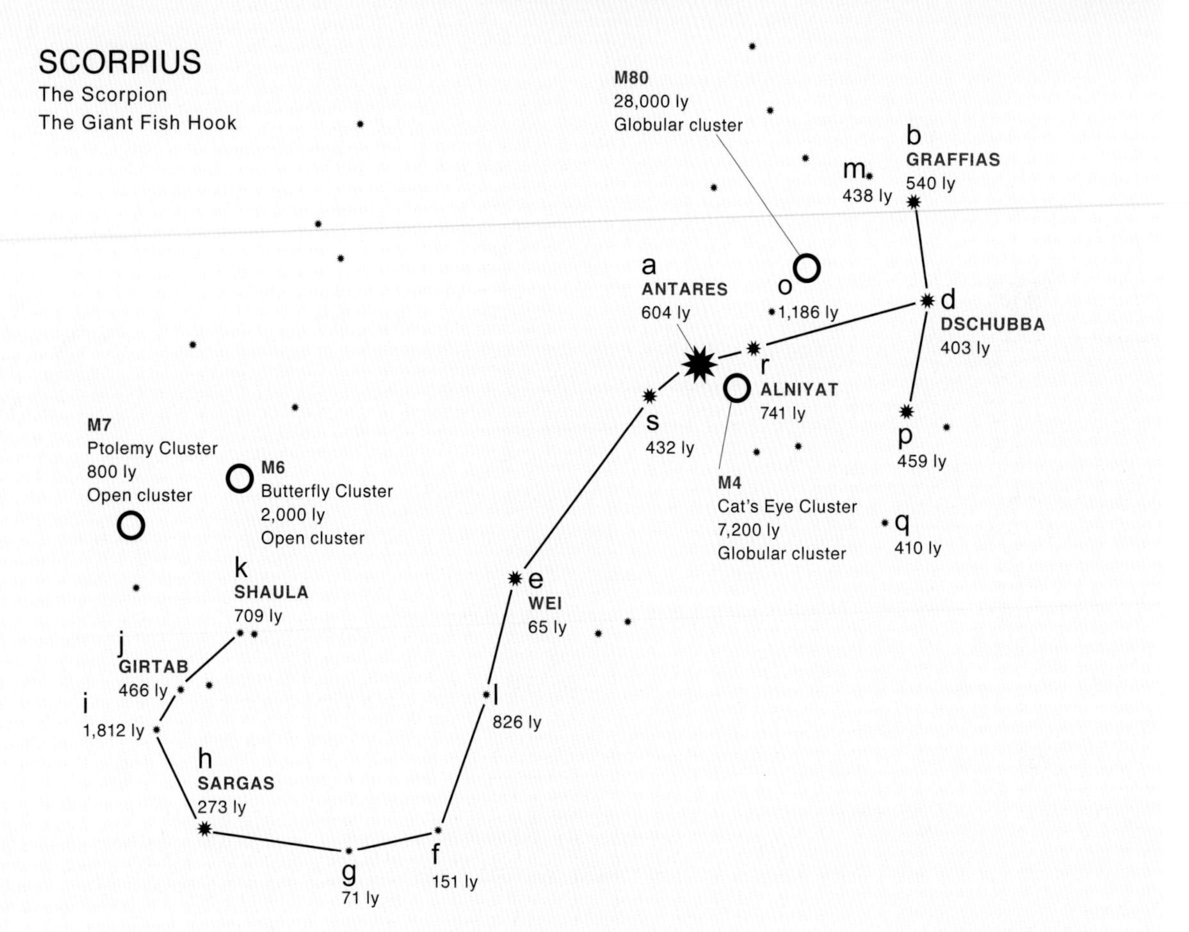

SAGITTARIUS THE ARCHER

Remember the kid's song that goes "I'm a little teapot, short and stout. Here is my handle. Here is my spout"? One of the classic summertime constellations, Sagittarius looks like a quaint teapot in the low southern sky, after evening twilight. You can easily find the handle on the left and the spout on the right. It even appears to pour its celestial brew on the tail of Scorpius the Scorpion, its neighbor to the west.

If you are a mythological purist, the constellation Sagittarius is a centaur shooting a bow and arrow. A centaur is a monster with a human head and the rear, tail, and legs of a horse. Lots of luck seeing Sagittarius as a bow-wielding centaur. You're better off looking for the teapot.

Not only is Sagittarius a cute constellation, it's in an important part of the sky in the foreground of the center of our galaxy. All the stars we see in the night sky are part of the Milky Way, which at last count may have a trillion stars in the shape of a giant spiral 100,000 light years across. When we see that collective glow of stars running north to south in the dark summer skies, we are looking edgewise into the plane of our galaxy. The glow around Sagittarius is brighter because that's the general direction of the Milky Way's center. It would be almost as bright as a full moon, but dark interstellar clouds and dust block the light from downtown Milky Way.

Since it's in the direction of the Milky Way's central region, Sagittarius is littered with celestial goodies. However, because Sagittarius never gets

THE FAR-FLUNG SCORPION

You can read about Scorpius in the Greek mythological tale of Orion. It was the scorpion sent by Zeus, the king of the gods, to fatally sting Orion, the mighty hunter. Orion had been having a nightly love affair with Artemis, the goddess of the moon. Zeus was not pleased to hear that his daughter was slacking on her job and fooling around with a mortal, so he paid Scorpius to kill the hunter in his sleep. Zeus's evil plan succeeded, and the following night Artemis saw the giant scorpion deliver the fatal sting. Artemis acted quickly. She picked up the fleeing scorpion and flung it into the night sky, where it magically transformed into a constellation. She then picked up Orion's body and gently placed him in the opposite end of the heavens. She wanted her boyfriend to keep her company in the sky as she guided the moon, and she wanted the giant scorpion as far away as possible. That is why we see the constellation Orion in the winter and Scorpius in the summer, prowling the southern sky.

high in our sky, we're forced to look through blurring atmosphere. It's still worth gazing at, though! Use a telescope or binoculars to see some great star clusters and nebulae; some you can even see with the naked eye.

Celestial Goodies

M8: The famous Lagoon Nebula is the second-best emission nebula in the sky, after the Great Orion Nebula. Whether you stargaze with the naked eye, binoculars, or a telescope, you can really see this beauty. This bright star factory is nearly 100 light years in diameter and cranking out new stars. Some astronomers estimate that its high levels of hydrogen gas could produce over a thousand new stars. A must see!
Magnitude: 5.0

M20: The Trifid Nebula is just north of the Lagoon Nebula and dimmer, but you'll like what you see with a small to moderate telescope. It looks like three separate nebulae but it's actually one large emission nebula with rifts of dark nebulae cutting across the foreground. Another must see!
Magnitude: 5.0

M21: This extremely young cluster contains about fifty stars shoved into an area about 10 light years wide. It's said to be 5 million years old, equivalent to a human baby only a few minutes old.
Magnitude: 7.0

M23: This is another great open cluster of about 100 stars squeezed into a diameter of about 15 light years. In our sky, it occupies an area the size of a full moon.
Magnitude: 6.0

M25: This loose open cluster contains about 100 stars in an area 20 light years across. In our sky, it's also about 1 moon-length in diameter.
Magnitude: 4.9

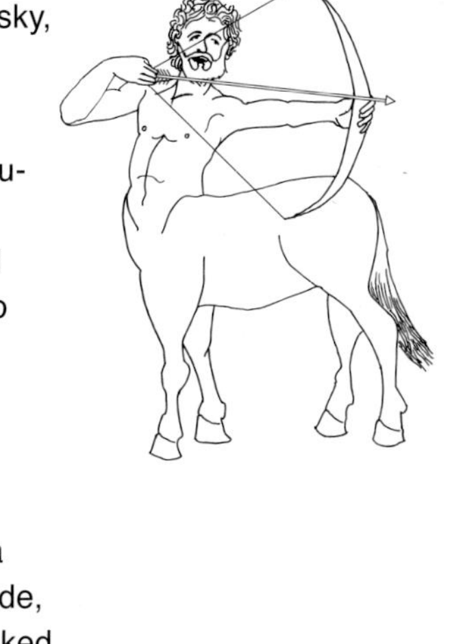

M22: The Sagittarius Cluster is a wonderful globular cluster easily seen with a small to moderate telescope. This globular cluster is packed with over a half-million stars shoe-horned into a sphere only 50 light years in diameter. It's another must see.
Magnitude: 6.5

M17: The Swan Nebula is a delightful emission nebula best viewed in dark, clear skies with a small to moderate telescope. In the countryside, you can spot it with binoculars or even the naked eye.
Magnitude: 7.0

M16: The Eagle Nebula is a bright nebula that contains an open star cluster. With clear skies, you can see a profile of a flying eagle. Of course,

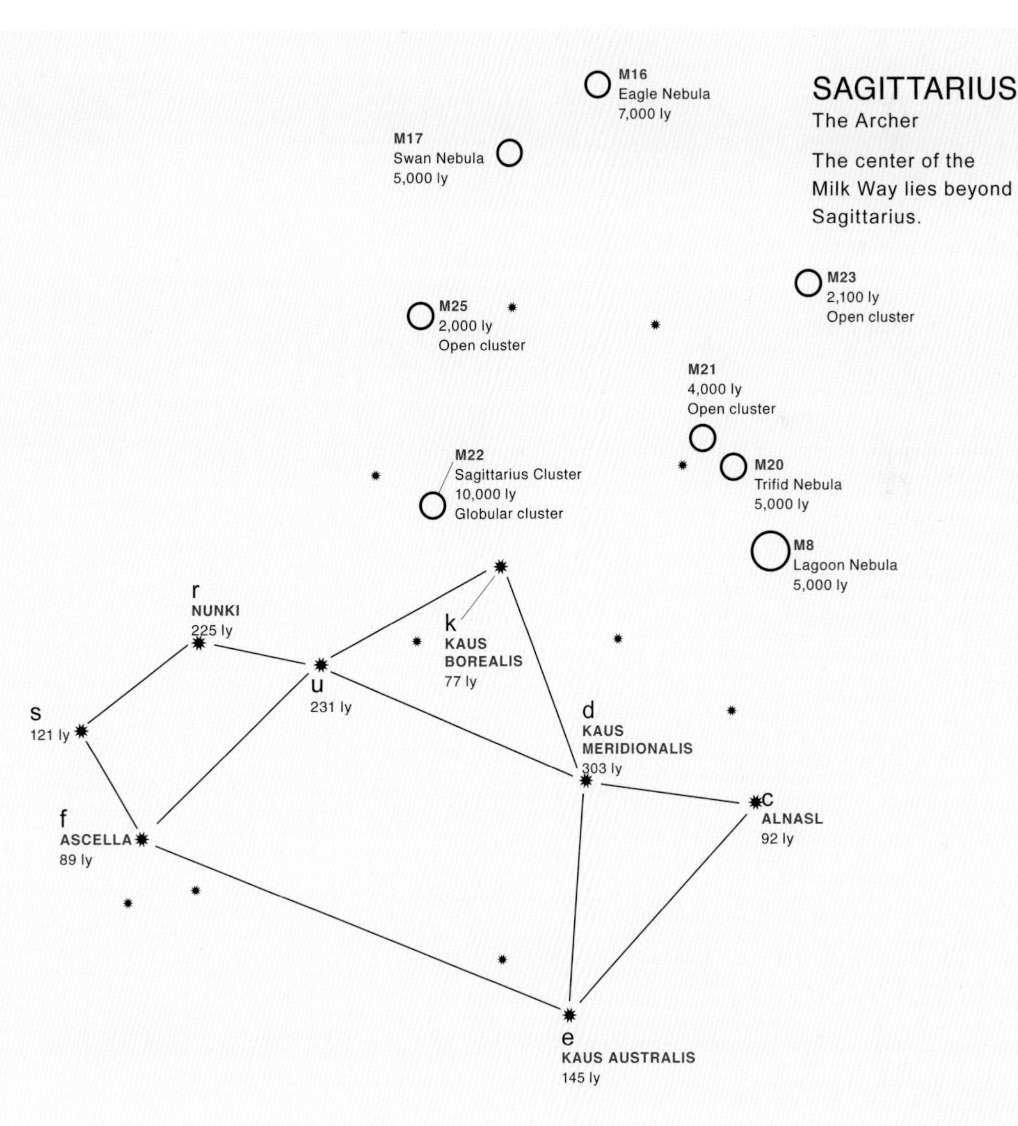

LEFT: Swan Nebula, M17 (Photograph © Russell Croman)

FACING PAGE: Lagoon and Trifid Nebulae, M8 and M20 (Photograph © Thomas Matheson)

the bigger the telescope, the better the likeness. The Hubble Telescope took a great photograph of the Eagle Nebula that has been nicknamed Fingers of Creation, so called because you can see the new stars being born.
Magnitude: 6.5

THE SUFFERING CENTAUR

If the teapot comes up short for you, the Greeks have a good tale about Chiron, the centaur. The centaur's father was a god, Cronus, who had many enemies and was forced to hide in the form of a stallion. When his son Chiron was born, he turned out to be half-human and half-horse. Chiron grew to be a skilled hunter and excellent marksman with a bow and arrow. He even taught young hunters the skills necessary to guarantee a hunk of meat on the supper table every night.

One of Chiron's students was the mighty hunter Hercules. One day, after extensive training, Chiron headed home while Hercules stayed behind to practice. As Hercules turned toward his paper target, one of his arrows slipped out and got caught in a strong wind. The misfired arrow pierced the departing centaur's back.

Chiron was badly wounded. If he were a mortal, he would have died. But being part god, Chiron lived on in excruciating pain. The pain was so bad that Chiron pleaded with the king of the gods to let him die and end his misery. Zeus denied the request at first, but finally took pity on the suffering centaur and granted him death.

Chiron's many godly friends honored him by flinging his body into the sky. They even tossed up his bow and arrow. We see him today as the constellation Sagittarius, visiting us every summer!

11

THE MARVELOUS BUT MISCHIEVOUS MOON

There's nothing like a moonlit night. In autumn, the classic harvest moon rises among the changing colors of the trees, bathing high-school football games and bonfires in moonlight. In winter, nothing is more beautiful than moonlight bouncing off a field of new snow. The summer moon rises over your campsite after a great day of swimming, fishing, or just taking it easy in the shade with a tall cool one.

Those are the good sides of the moon, but as any stargazer will tell you, the moon is the enemy of stargazing, especially when it's full. The moon's evil light washes out all but the brightest stars and constellations. A lot of amateur astronomers, including myself, stow away their telescopes during a full moon. You can't go skating in a buffalo herd, and you can't stargaze under a full moon!

A full moon (Photograph © Richard Hamilton Smith)

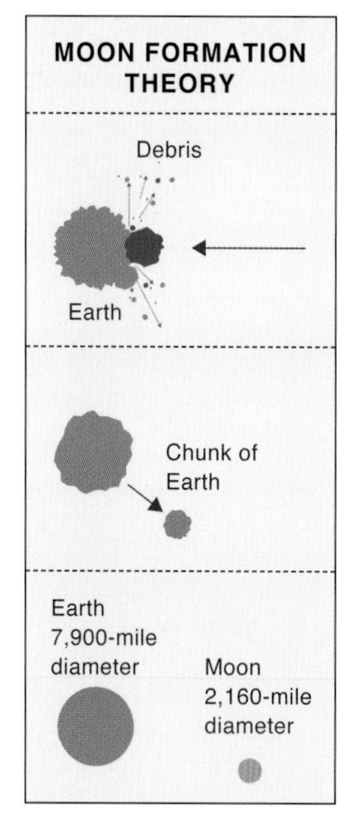

TOP: The moon as seen from the spacecraft Galileo (Courtesy of NASA)

All right, I might be a little tough on the 4.5-billion-year-old moon. It's actually a lot of fun to look at the moon though a telescope, especially knowing that back in the late 1960s and early 1970s, twelve men walked on the lunar surface. So far, though, no one has spotted the lower landing platforms left by the lunar modules, or Richard Nixon's signature on the commemorative plaque.

If you were alive in 1969, you know exactly where you were when Apollo 11 landed that Sunday afternoon, July 20. I was at CYC Camp on Big Sandy Lake near McGregor, Minnesota, watching Walter Cronkite on the head counselor's twelve-inch black-and-white TV yell, "Man on the moon!" That experience helped hook me on astronomy and stargazing. That day, Neil Armstrong became the first person to step on the moon, with the words, "One small step for man, one giant leap for mankind." His partner, Edwin "Buzz" Aldrin, stepped on the lunar surface a few minutes later. It wasn't published back then, but when Buzz strolled on the Sea of Tranquility, he walked on water . . . kind of. The urine storage bag in his boot burst when he descended the ladder of the lunar module. When he bounced on the moon in one sixth of the earth's gravity, he sloshed around in his own pee. Somehow, I think the excitement of being the second human on the moon compensated for this indignity.

After Apollo 11, ten other men walked on the moon. Do you know the name of the last man to strut on the lunar surface? I'll give you a hint: it was in December 1972. The answer is at the end of this chapter.

WHERE DID THE MOON COME FROM?

There are several theories about where the moon came from. One theory is that the moon and the earth formed together out of the same clouds of gas and dust that eventually became our solar system. Another is the "captured moon theory," which argues that the moon was a captured asteroid or other debris from the formation of the solar system.

The leading theory is that our moon formed toward the end of the solar system's formation, when a lot of debris was flying around. A large object, about twice the mass of Mars, collided with early molten Earth and knocked a chunk out of it. The earthen chunk cooled and became our moon. As the moon cooled, it took on the spherical shape we see today. It wound up being one-quarter of the earth's size, 2,160 miles in diameter, compared with the earth's 7,900-mile girth.

THE SURFACE OF THE MOON

When you gaze at the moon, even without a telescope, you see black splotches covering large portions of the surface. Galileo and other seventeenth-century astronomers, looking through the newly invented telescope, decided these splotches were large bodies of water. They called them maria, which means seas. That's why moon maps refer to these areas as the Sea of Tranquility, the Sea of Crisis, and the Ocean of Storms. Even though today we know maria are not lunar oceans, the name has stuck. Maria are actually large expanses of dark, smooth lowland lava plains. Huge meteor impacts during the formation of the solar system cracked the lunar surface, causing massive lava flows that cooled and became the maria. These flat areas were the landing sites for the first Apollo landings. In later missions, the landing sites were more daring, with spacecraft descending on more mountainous areas.

Along with maria, the moon's surface is covered with thousands of craters, some of which you can see with the naked eye. One prominent crater, visible especially when the moon is full, is Tycho, on the southern edge of the moon's disk. It's over 50 miles wide, and you can see rays spreading in all directions from the crater. Tycho is thought to have formed from a giant meteor crashing into the moon. The rays, ejecta from the crash, reflect sunlight better than the surrounding surface because the spewed material isn't as old. Tycho is considered a young crater, only about a billion years old.

Through a telescope, you can see many more craters, thousands and thousands of them, all different shapes and sizes. Most are nearly as old as the moon itself. They formed during the final phases of the solar system's formation, over 4 billion years ago. At the time, giant meteors and other projectiles were flying around and colliding with the moon and the earth. The moon, not having much of an atmosphere to absorb meteors, was at the mercy of this flying debris. Meteor collisions formed most of the craters we see today. The moon's atmosphere was a result of its weak gravitational pull, which was a result of its small mass, 1/81 the mass of the earth. The earth, despite its atmosphere, got its share of impacts, but erosion has weathered away most of its ancient craters.

When astronauts walked on the moon, they wore big, bulky space suits. Since the moon has no

A moonwalker
(Photograph © Comstock)

Earth and its moon (Courtesy of NASA/JPL-Caltech)

atmosphere, it has no air, and the surface temperature fluctuates between extremes. In the sunlight, the temperature can be over 230°F, and in the shadows, it can be colder than 300°F below zero! It's no resort up there!

THE MOON'S ORBIT

Why does the moon change shape? Why is it a crescent one night, a half-moon another night, and full another night? Why do we not see the moon at all some nights? The moon's changes are determined by two factors: sunlight reflecting off the moon and the moon's orbit around the earth.

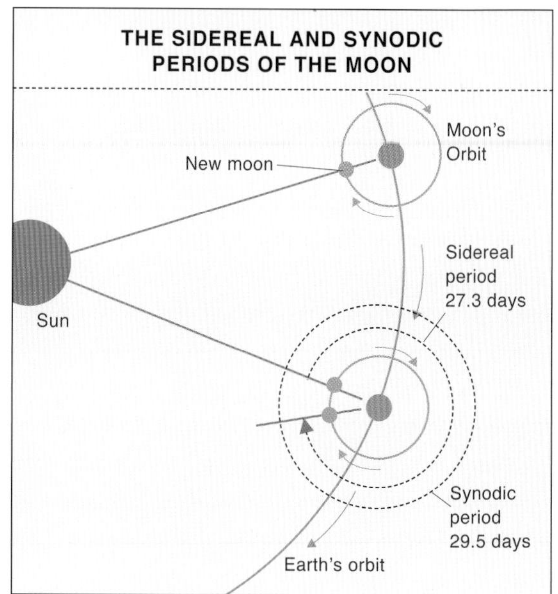

The sun is the only object in our solar system that generates light. The light we see on a moonlit evening is reflected sunlight. As the moon revolves around the earth, we see different lunar phases due to the changing angle between the moon, the earth, and the sun. The moon makes a full revolution every 27.3 days in what's known as the sidereal period.

The lunar orbit is elliptical, or lopsided. Because of that, the moon's distance from us varies. Its farthest point, called aphelion, is over 252,000 miles away; its closest point, called perihelion, is around 222,000 miles away. The changing distance makes some full moons larger than others. In fact, when a full moon occurs in perihelion, it can be over 30 percent brighter and 14 percent larger than average.

Don't confuse this phenomenon with the larger appearance of the moon when it's rising or setting. That size change is just an optical illusion that happens because you're comparing the moon with land objects close to the horizon. To make that illusion go away, here's what you do, assuming you're physically able to: moon the moon! This is no "loony" joke. Face away from the moon, bend over, and look at the rising or setting moon between your legs. That will make the moon look its "normal" size, when it's high in the sky.

By the way, the moon and sun glow orange-red near the horizon because the earth's atmosphere scatters away all but the reddish components of light. Even stars and planets appear more ruddy when they are close to the horizon.

PHASES OF THE MOON

The best way to explain the phases of the moon is to look at the different positions in its orbit.

New Moon

Every 29.5 days, the moon lies roughly in line between the earth and the sun, which means the sun illuminates only the side of the moon that faces away from the earth. From our perspective, the moon is invisible, rising at sunrise and setting at sunset. This is called a new moon. Every once in a while, the moon lies exactly in line between some point on the earth and the sun, and we have a solar eclipse.

Waxing Crescent Moon

A few days after the new moon, the angle between the moon, earth, and sun opens up a bit and we begin to see a sliver, or crescent, of the sunlit part of the moon. The moon moves east of the sun in the sky each day, so the waxing crescent moon trails behind the sun a little more every evening. It rises shortly after sunrise and sets shortly after sunset, allowing us to see it for a few hours after twilight in the western sky. This is a great time to see a phenomenon called Earthshine. That's when you not only see the crescent of the moon, but you also see the rest of the moon bathed in much dimmer light. It's called Earthshine, but it's actually secondhand sunshine. Whenever the moon is in the crescent phase, the earth would be nearly full to an astronaut looking back from the moon. Some of that sunlight reflecting off the full earth shines off the darkened part of the moon and makes a lovely spectacle.

First Quarter Moon

Seven days after the new moon, we have a first quarter moon. It's called first quarter because the moon is one-quarter of the way through its cycle of phases. It doesn't mean that we only see a quarter of the moon. We actually see a half moon because half of the sunlit part of the moon faces the earth. The earth, moon, and sun are positioned at a right angle to each other. The moon rises around midday and sets around midnight. This is a wonderful time to view our lunar neighbor with a telescope—especially take a look along what's known as "the terminator," which is the line between the darkened part of the moon and the sunlit part. Along the terminator, the shadows are long, revealing features that are otherwise harder to see. You can even see the mountain peaks poking above the shadows on the dark side of the terminator. If you were standing on the moon somewhere along the terminator, you would see a lunar sunrise.

The lunar surface
(Courtesy of NASA Apollo)

Waxing Gibbous Moon

Ten days after the new moon, the angle between the moon, sun, and earth opens up beyond a right angle, and we see more of the sunlit half of the moon. Something that is gibbous is convex or rounded in shape; for the moon, this means it is more than half illuminated but not full. The growing moon takes on an oval shape and kicks too much light into the sky for decent stargazing. The waxing gibbous moon rises in the middle of the afternoon and sets around two or three o'clock in the morning.

The Full Moon

Fourteen days after the new moon is the full moon. We're now halfway through the moon's twenty-nine-day cycle. The moon is on the opposite side of the earth from the sun, and from Earth we see the complete sunlit half of the moon. The full moon rises at sunset and sets at sunrise, and while moongazing is wonderful for love and romance, stargazing is

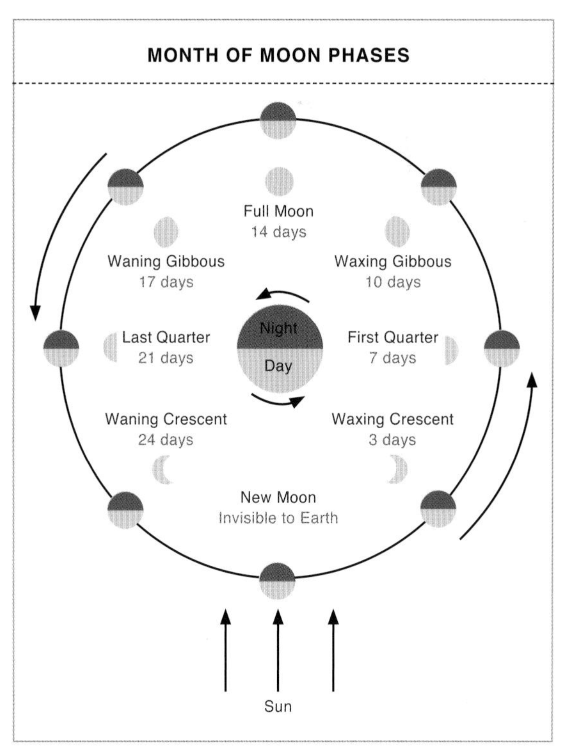

ABOVE, TOP: Waxing crescent moon with Earthshine. (Photograph © Thomas Matheson)

ABOVE, BOTTOM: First Quarter Moon (Photographer © Robert Gendler)

toast! In fact, this isn't a good time to explore the moon either, because everything on the moon's surface is in direct sunshine and there are no revealing shadows.

Around the time of the full moon, you see the classic "Man on the Moon." A lot of people, however, see a woman on the moon. In fact, I see Wilma Flintstone. It's also possible to see Mr. Bill from the TV classic *Saturday Night Live*, yelling "Nooooooooooo!" It's also possible to see the profile of JFK, a Scottie dog, and a rabbit—not all at the same time, of course!

Once or twice a year, the full moon enters the reddish shadow of the earth and we have a lunar eclipse.

Waning Gibbous Moon

About seventeen days after the new moon, you'll notice the moon shrinking as the angle between the moon, the sun, and the earth closes up. We see an oval moon again, the mirror image of the waxing gibbous. The waning gibbous moon rises after sunset and sets after sunrise. That's when you see the moon in the western sky in the morning after the sun is already up.

Last Quarter Moon

Twenty-one days after the new moon, we have a last quarter moon. Once again the moon, the sun, and the earth are positioned at a right angle. We see the opposite half of the moon from the half we saw during the first quarter. The last quarter moon rises about midnight and sets around noon. Just like the waning gibbous moon, we see the last quarter moon in the western sky after sunrise.

Waning Crescent Moon

About twenty-four days after its new phase, the moon shrinks to a crescent again as the angle between the sun, the moon, and the earth gets smaller and smaller. The waning crescent moon rises two to three hours before sunrise and sets in the early afternoon. It's so close to the sun that it becomes invisible shortly after sunrise. Before sunrise, through, you have a chance once again to see Earthshine with the waning crescent moon.

Back to New Moon

Twenty-nine-and-a-half days later, the moon is once again new and the whole phase cycle, called the synodic month, starts all over again. As the moon orbits the earth, not only does it change shape, but it migrates eastward among the stars about 13 degrees, or twenty-six moon-widths, every twenty-four hours. Because of that migration, it rises twenty to fifty minutes later each night, depending on the time of year.

If you're a keen observer, you've noticed that the synodic period, or month of phases, is 2.2 days longer than the sidereal month, the actual orbit period of the moon around the earth. This is because while the moon revolves around the earth, both the earth and the moon revolve much more slowly around the sun. The synodic period is longer because the moon has to "catch up" to once again be on line between the sun and the earth to reach another new moon.

SOLAR ECLIPSES

One of the biggest celestial coincidences of nature is the fact that the sun's disk in our sky appears to be just about the same size as the moon's disk. Even though the sun's diameter is 390 times that of the moon, the sun is also 390 times farther from the earth. Because of that coincidence, we get to experience the rare beauty of total solar eclipses. When I say rare, I mean rare. On average, for any one spot on the earth, a solar eclipse occurs once every 360 years with totality lasting only a few minutes. A fairly recent total eclipse occurred February 26, 1979, in Winnipeg, Manitoba—I know because I was there with my little brother Jim.

Remember the lyric in the song "You're So Vain" by Carly Simon, "You flew your Lear jet up to Nova Scotia to see the total eclipse of the sun"? In my case, I drove my rusted 1969 Ford Galaxy up to Winnipeg to see something I'll never forget. Starting from the University of Wisconsin at Madison, I drove north, picked up my brother at home in Richfield, Minnesota, and flew up the freeway. Words can't adequately describe the magic and beauty of a total solar eclipse. As totality approached, the temperature dropped, weird shadows appeared, and the horizon in all directions took on a strange-looking pink twilight. Just before totality, some stars came out and I saw the Big Dipper.

During the couple of minutes of totality, I was able to safely look through my telescope at the eclipsed moon. I'll never forget the solar flares shooting away from the sun or the ghostly white ring of the sun's corona. Believe me, if you have a chance to see a total solar eclipse sometime in your life, do it . . . and pray for clear skies! On average, an eclipse occurs somewhere on the earth every year and a half. The next total solar eclipse in the United States will be August 21, 2017. The path of totality will be from Georgia to Nebraska to Oregon. Take care of yourself so you'll be around for it!

A total eclipse is rare because the moon has to be in exactly the right place at the right time. It can only happen when the moon is new, but most new moons never result in a solar eclipse. That's because the moon's orbit around the sun is inclined by five degrees to the earth's path around the sun. During the vast majority of new moons, the moon passes above or below the plane of the earth's orbit. But when the new moon finds itself on the plane of the earth's orbit, a total eclipse traces a

BELOW, TOP: Waning gibbous moon (Photograph © Robert Gendler)

BELOW, BOTTOM: Total solar eclipse (Photograph © Jack Newton)

The diamond ring effect during a solar eclipse (Photograph © Jack Newton)

path of totality somewhere across a small patch of the earth. There's a path because, while the moon covers the sun, the earth continues to rotate under the spectacle.

You don't need to protect your eyes during the totality part of a eclipse, but if the moon is just partially covering the sun, you must not look at it directly! You can do permanent damage to your eyes. It's even worse if you look at the sun through any size telescope. Your retina would instantly burn and you would go blind in the eye, permanently! The safest way to look at a partial eclipse is to project the image of the eclipse through a pinhole in a piece of cardboard onto another piece of white cardboard. You can also look at it through extremely thick #14 welder's glass.

Some solar eclipses don't quite cover up the entire sun. That happens when the moon is at apogee, its farthest point away from the earth. These kinds of eclipses are called annular solar eclipses. During the totality of an annular eclipse, a ring of the sun's surface surrounds the eclipsed moon.

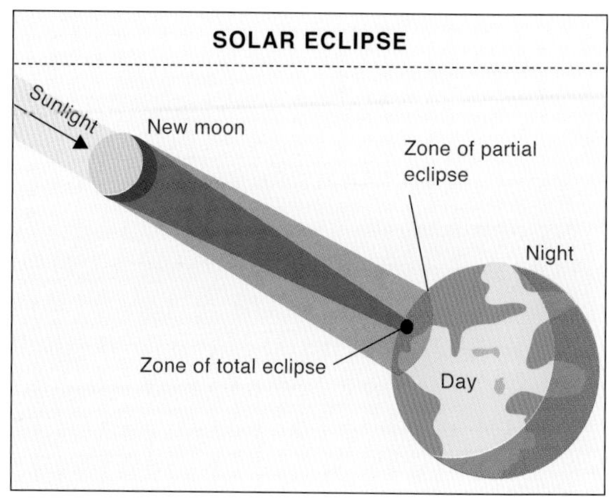

We're living in a privileged time to be able to see total solar eclipses, because the moon and sun's disks are nearly the same size in the sky. Starting around the year A.D. 600,000,000, all total solar eclipses will be annular. That's because the moon is slowly drifting away from the earth at a whopping 1½ inches per year. Bye bye, moon!

LUNAR ECLIPSES

Lunar eclipses aren't as spectacular as solar eclipses, but the trade off is that they are more common. One to two lunar eclipses occur every year on average, and anybody on the nighttime side of the earth can see it. You don't have to ride that Lear jet anywhere.

Lunar eclipses occur when a full moon enters the earth's umbra, which is a dark shadow that cuts off all light. The 5-degree difference between the moon's orbit around the earth and the earth's orbit around the sun accounts for why the full moon misses the shadow most months. The full moon doesn't have to be exactly in line for a lunar eclipse, because at the moon's distance, the umbra is a little over 6,000 miles wide. Because of the wide shadow, totality in a lunar eclipse can last over an hour, compared with totality in a solar eclipse that only lasts a few minutes.

During a lunar eclipse, the moon doesn't go dark. Instead, it goes red from filtered sunshine making its way through the surrounding shell of the earth's atmosphere. Our atmosphere does two things during a lunar eclipse. It filters all but the red components of the sun's light, and it bends that red light toward the direction of the moon. The red light we see on the moon is actually the combined twilight of all of the sunrises and sunsets happening on the earth during the time of the eclipse. The degree of redness depends on how deeply the moon has entered the umbra. It can also be affected by pollution. Lunar eclipses that occur after a large volcanic eruption yield especially dark shades of red.

You can look at lunar eclipses as long as you want and with whatever you want—the naked eye, binoculars, telescopes, or whatever. Outside of the possibility of becoming a little loony, nothing bad will happen to you. One phenomenon to watch for with a telescope is the moon passing in front of, or eclipsing, background stars. This is called occultation. Since the moon doesn't have an atmosphere to speak of, the stars blink in and out instantly. Now you see them; now you don't.

By the way, the last person to walk on the moon was Apollo 17 astronaut Gene Cernan!

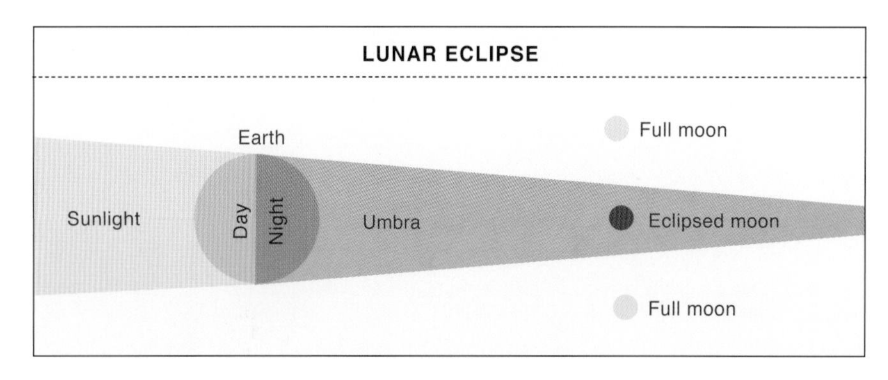

Lunar eclipse (Photograph © Thomas Matheson)

12

PLANETS, WANDERERS OF THE SKY

When you gaze into the night sky, how can you tell a planet from a star? Planets generally appear brighter, and they don't twinkle. Planet light looks steadier than starlight because planets are closer to the earth than are the stars. And while our atmosphere jostles incoming light waves, making stars seem to twinkle, planets give off so many light rays that their light stays steady. One exception is planets close to the horizon; their light pierces through so much atmosphere that they occasionally twinkle. Remember that the light you see from planets is reflected sunlight. Planets generate no energy or light of their own—unless you count the glow shopping centers send out from Earth.

Another way to tell planets from stars is to remember that planets move among the stars. In fact, the word planet comes from the Greek word *planetes*, which means wanderer. Back in ancient Greece, no one knew what the planets were. All they knew was that these bright objects wandered across the celestial sphere, among the stars within a band of twelve constellations now known as the zodiac constellations.

This montage shows the phases of Saturn. The planet's rings are tilted to its orbital path, making the rings appear at different angles from Earth. (Courtesy of NASA and the Hubble Heritage Team (STScI/AURA))

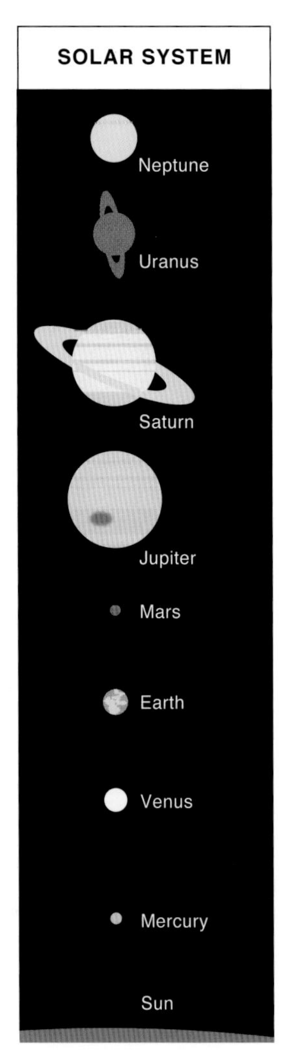

We know a whole lot more about planets nowadays, but just like our ancestors, we still see them from our backyards as bright star-like objects wandering amongst the background of "fixed" stars. Planets migrate at various speeds, generally eastward against the background stars, but at times they take a westward course. Mercury and Venus, which are known as inferior planets because their solar orbits lie inside the earth's orbit, have regular eastward and westward motions. The other planets are all superior, not because they are more important than Venus or Mercury, but because they have orbits that lie outside the earth's solar circuit. The superior planets tend to move eastward against the background stars.

RIGHT: This montage of our solar system forefronts the most distant planets and trails back to the sun. (Courtesy of NASA)

FACING PAGE: Our beautiful Earth (Photograph © Comstock)

MOVEMENT OF THE INFERIOR PLANETS

Since Mercury and Venus orbit the sun within the earth's orbit, neither gets far from the sun in our sky. Mercury in particular is a sun hugger, preventing us from getting a good look at it. If you see Mercury, it's only for a brief time shortly before sunrise or shortly after sunset. It's usually bathed in twilight. Venus's orbit is a little farther from the sun, so we see it for longer periods of time before dawn or after dusk, but it still doesn't stray all that far.

To understand the motion of the inferior planets in our sky, let's examine one of the planet's orbits in detail. We'll look at Venus, since its orbit is larger. If we trace Venus's orbit from our vantage point on Earth, we see a fairly straight line extending east and west of the sun, because the earth and Venus orbit the sun in nearly the same plane. Both the inferior planets move equally east and west against the background stars.

Every 584 days, Venus lies roughly in line between the earth and the sun, placing it at its closest point to the earth, called inferior conjunction. On rare occasions, Venus actually passes in front of the sun's face and we have what's known as a transit.

Seventy-two days after inferior conjunction, Venus moves westward among the stars to what's known as its greatest western elongation. In those seventy-two days, Venus is a "morning star," and we see a crescent shaped Venus (just like a crescent moon) rise earlier and earlier before sunrise. At greatest western elongation, 48 degrees west of the sun, Venus rises three to four hours before sunrise, giving us a half-Venus. After greatest western elongation, Venus migrates eastward, showing more of its face but also getting farther away from us. It takes 220 days to go from greatest western elongation to superior conjunction behind the sun, where it's out of our view and at its maximum distance.

After superior conjunction, Venus continues eastward for another 220 days, until it reaches greatest eastward elongation, at which time Venus sets later and later after sunset and becomes an "evening star." After greatest eastern elongation, Venus resumes its westward motion for seventy-two days until it arrives back at inferior conjunction. During this period of western movement, Venus sets earlier and earlier after sunset, and we see a shrinking crescent through the telescope.

The cycle of Venus's motions in our sky is called the synodic period. The planet's orbital period around the sun, called the sidereal period, equals 225 Earth days. Why is there a difference? While we watch Venus go around the sun, the earth is also orbiting the sun at a slower speed. That's why Venus moves faster among the stars when migrating west and slower when going east.

FACING PAGE: The planets Venus and Jupiter are visible as "stars" below a crescent moon. (Photograph © Russell Croman)

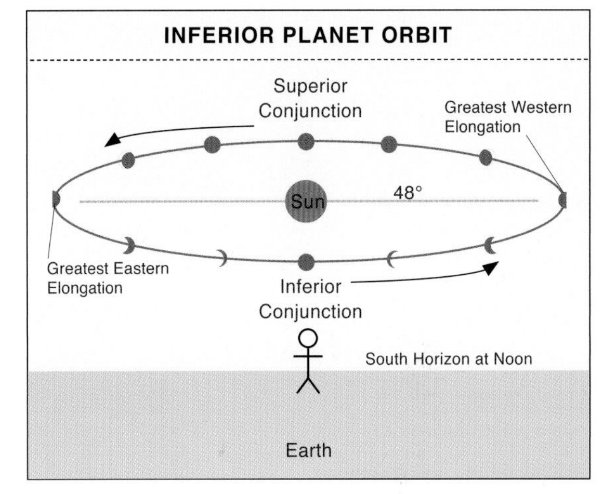

You see the same kind of motion for Mercury as for Venus, and, as expected, the orbital times are smaller and the elongations shorter.

MERCURY

Named after the messenger of the gods, Mercury is the closest planet to the sun, just 36 million miles from our home star. It orbits the sun faster than any other planet, completing a cycle in eighty-eight days. Mercury also rotates once on its axis every fifty-nine days. At only 3,032 miles in diameter, Mercury is not much bigger than our moon. In fact, with all its craters and its absence of clouds or an atmosphere, it looks just like the moon. It's hotter, though, with the temperature in the blasting sunshine about 700°F. On the nighttime side of Mercury, the temperature plummets to 300°F below zero. Some astronomers believe that, just like our moon, the polar regions contain water permanently hidden in deep craters, away from sunshine.

Mercury (Courtesy of NASA/JPL-Caltech)

A highly protected astronaut on Mercury would see the sun's disk as three times larger in the sky than we see it. I recommend at least a 385-SPF sunscreen!

VENUS

If Mercury's 700°F temperature isn't hot enough for you, try spending a weekend on Venus. You won't need that extra sweatshirt with surface

temperatures around 900°F, the hottest in the solar system. If you could take that kind of heat, you'd also have to put up with massive quantities of carbon dioxide, which is poisonous to the lungs; thick cloud cover; and occasional sulfuric-acid rain showers. Finally, the atmospheric pressure—ninety times greater than on the earth—would crush your body!

Venus is similar in size to the earth, about 7,520 miles in diameter. Named after the goddess of love, Venus is anything but lovely. It's the true hellhole of our solar system, a greenhouse gone mad. The thick, heavy atmosphere, mostly carbon dioxide, lets in radiation from the sun but traps a great deal of terrestrial radiation emerging from the planet's surface. You end up with a poisonous pressure cooker!

ABOVE: Venus (Courtesy NASA/JPL-Caltech)

RIGHT: Mars (Photograph © Comstock)

A thick cloud cover hides the surface of Venus, but radar images taken from the Magellan spacecraft reveal mountains higher than those found on Earth and humongous valleys that dwarf the Grand Canyon. Venus also has active volcanoes that spew molten rock and poisonous gas into the atmosphere. I'll take a rainy weekend at the lake anytime!

TELESCOPING MERCURY AND VENUS

The trouble with viewing Mercury and Venus through a telescope is that you can't see much. Mercury is so small and close to the sun that all you get is a blurry image. Just spotting it is an accomplishment!

Venus is just as bad, even though it's bigger than Mercury and usually closer to the earth. Its cloud cover completely hides the surface. Sunshine bouncing off these clouds makes Venus, at times, the third-brightest celestial object in our sky, behind the sun and moon. Many a time Venus has been reported as a UFO.

MOVEMENT OF THE SUPERIOR PLANETS

Superior planets generally move east among the zodiac constellations. Their speed among the stars depends on how fast they move around the sun. For example, Mars, in its 687-day orbit, moves more rapidly through the zodiac than Saturn, which takes twenty-nine years to make one solar circuit.

The visibility cycle of superior planets is simpler than the visibility cycle of the inferior planets. I'll use Jupiter as an example. When the earth is in line between the sun and Jupiter, Jupiter is said to be in opposition, because it's on the opposite side of the sky from the sun. This is the best time to see a superior planet because not only is it at its minimum distance from the earth, it's also out all night. Just like a full moon, it rises at sunset and sets at sunrise.

As Jupiter and Earth continue their orbits at their individual speeds, they eventually arrive at a point where the sun lies in a line between Earth and Jupiter. This point is called superior conjunction, and it is the worst time to go Jupiter hunting. Not only is the planet at its maximum distance from us, it's behind the sun.

The time from one opposition to the next is called the synodic period, and for Jupiter that's 399 days, a little over thirteen months. In the one year it takes the earth to revolve around the sun, Jupiter only moves about 1⁄12 of its orbit. It takes the earth just over another month to catch up to another opposition.

In the case of Saturn, the time between oppositions is a little less: 378 days. While the earth makes one revolution around the sun, Saturn only creeps along 1⁄29 of its solar orbit. On the other hand, Earth and Mars constantly play a much longer game of cat and mouse. During the earth's yearly orbit, Mars moves about halfway around the sun, so the earth needs a lot more time to catch up to another opposition. The synodic period of Mars is 780 days, which means we have a close encounter with Mars every 26 months. Also, because of the eccentricity of Mars's orbit, the minimum distance at oppositions varies. In August 2003, Mars and Earth were about 34 million miles apart, the closest we've been to the red planet in almost 60,000 years!

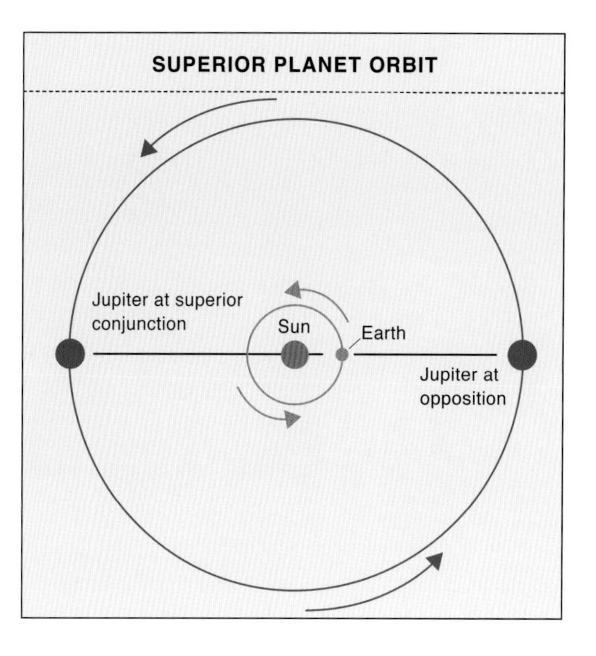

Around the time of opposition, superior planets retrograde westward as the speedier Earth whizzes past them. Also, unlike inferior planets, which change shape, superior planets more or less show us their entire sunlit side.

MARS

Named after the god of war, this blood-red planet has always been a source of great interest. It seems the more we discover about Mars, the more we question. Ever since Italian astronomer Schiaparelli thought he saw canals on the red planet, we have been fascinated with the possibility of life on Mars. The canals proved to be an illusion, but we still wonder about everything from little green men to microbes.

Truthfully, Mars would be a tough place for life to exist. First, its color has nothing to do with blood. The planet's red hue comes from iron oxides, or rust, on its surface. It has a thin atmosphere, only 1 percent of Earth's atmosphere, and 90 percent of that is carbon dioxide. The Mars landscape features craters, deep canyons as large as the continental United States, and huge dormant volcanoes. It also experiences strong winds that cause global dust storms. Temperatures range from 250°F below zero in darkness to 90°F above zero in sunshine. It has

two polar caps that contain some water, but the caps are mostly frozen carbon dioxide.

No signs indicate any other current surface water, although all kinds of visible evidence points to water in the planet's distant past. Telescopes and orbiting spacecraft have photographed dried river valleys, lakebeds, and flooded plains. All this surface water is believed to have evaporated billions of years ago when Mars lost its atmosphere.

In the last few years, an exciting discovery was made. The Mars Odyssey orbiter detected ice reservoirs below the surface, all over the planet. In fact, it's believed that if all this ice was brought to Mars's surface, the entire 4,000-mile-wide planet would be covered by ankle-deep water.

In 1996, astronomers made another exciting discovery, when they found a meteorite in Antarctica that had bounced off Mars and found its way to Earth. Thought to be over 4.5 billion years old, the meteorite showed signs of micro-bacteria. Whether or not this "life" came from Mars is still under debate. Stay tuned!

TELESCOPING MARS

Most of the time, all you see of Mars, even through the biggest telescopes, is a small pale-red disk. But during opposition, about every two years, focus your scope on the red planet and see dark markings that are Mars's rocky plains and valleys. You can also see at least one of its white polar caps. Since Mars rotates on its axis once every 24.5 hours, the face of Mars always changes.

Be patient with Mars. It's tricky. Some nights it's clearer than others. This enigmatic planet is susceptible to atmospheric turbulence. Also, make sure it's more than 25 degrees above the horizon so you don't have to look through as much of the earth's atmosphere.

JUPITER

Even though it's essentially a big ball of hydrogen gas half a billion miles from the sun, Jupiter is the king of the planets! It is twice the mass of all the other planets in our solar system combined. If Jupiter were hollow, you could fill it with over 1,300 Earths. It's over 88,000 miles in diameter and even fatter at the equator because it spins on its axis at a dizzying rate, rotating once every nine hours and fifty-five minutes.

We know a lot about Jupiter from two Voyageur spacecrafts that passed it in the late 1970s, and the Galileo spacecraft that explored Jupiter and its moons in the 1990s. Jupiter has no real surface, just denser and denser gas layers toward the center of

FACING PAGE, TOP: Mars passes by the moon as the red planet approaches opposition. (Photograph © Thomas Matheson)

FACING PAGE BOTTOM: The surface of Mars as viewed from the Mars Exploration Rover Spirit in 2004 (Courtesy of NASA/JPL/Cornell)

ABOVE: Jupiter's Red Spot (Courtesy of NASA/JPL-Caltech)

LEFT: Jupiter (Courtesy of NASA/JPL/Space Science Institute)

TOP: Saturn (Courtesy of NASA and the Hubble Heritage Team (STScI/AURA))

ABOVE: Io, the Pizza Moon (Courtesy of NASA/JPL)

the planet. Eventually it becomes liquid hydrogen, and then metallic hydrogen. Jupiter is also thought to have a rocky core.

Pictures of Jupiter show colorful horizontal cloud bands with embedded storms, including the Great Red Spot, so big that you could line three Earths across it. The colors represent hydrogen and helium mixed with methane, sulfur, ammonia, and other gases. These clouds and storms whip around Jupiter at hundreds of miles an hour. Astronomers think the constant gravitational collapse of Jupiter's gases produces this energy. Jupiter's atmosphere also produces lots of lightning. In the Red Spot, the lightning is more than a billion times more powerful than lightning on Earth.

Jupiter's four biggest moons were first seen in the early 1600s by Italian astronomer Galileo Galilei. Using a crude telescope, Galileo observed how the moons Io, Europa, Gaymede, and Callisto changed positions as they circled Jupiter. Galileo used this evidence to support Nicolaus Copernicus's sun-centered theory of the solar system. His conclusions got Galileo in trouble with the Catholic Church, which was not pleased to hear someone say Earth and humans weren't the center of the universe. The Inquisition eventually placed Galileo under house arrest for heresy. It was a dark time for astronomy. It took almost 400 years, but Pope John Paul II finally cleared Galileo of any charges.

Io and Europa are the most fascinating of Jupiter's Galilean moons. Volcanoes constantly erupt on Io, the closest moon to Jupiter, which is a little smaller than our moon. Immense tidal forces from Jupiter keep the inside of Io hot and explosive. So much lava flows on Io that NASA refers to it as the Pizza Moon. The surface of Europa, on the other hand, appears to be a layer of ten-mile-thick ice. Below the ice may be liquid or slushy water. Could Jupiter's tidal forces heat this water enough to support life? Who knows, some day ice-fishing houses could dot Europa.

TELESCOPING JUPITER

Through just about any telescope, you can see the disk of Jupiter and some of its cloud bands, especially the two dark bands on either side of the equator. You may also see the famous Red Spot, which looks more like a pink spot. You don't always see it, sometimes because turbulence in our atmosphere blurs the view, and other times because the Red Spot is on the other side of Jupiter.

Telescopes also offer good views of the four big Galilean moons dancing around Jupiter in periods of two to seventeen days. My buddy Dave Lee can even see them around opposition with his eagle eyes.

The moons look like stars lined up on either side of Jupiter. Because of their orbits, they are in different alignments every evening. You won't always see all four since one or more may be in front of or behind the planet. If your telescope is powerful enough, you can see the shadow of a moon cross in front of the king of planets.

SATURN

Saturn is the crown jewel of the solar system. Whenever my stargazing classes turn their scopes on Saturn, I hear all kinds of superlatives. I know I've hit a homerun when one of the kids says "sweet."

Mention Saturn to the average person and they say, "That's the planet with the rings." The truth is all the giant gas planets in our solar system have rings, but Saturn's are the most vivid. When Galileo saw Saturn through the first telescope, he thought he saw a planet with cup handles.

Saturn, a little less than a billion miles away, is made up of hydrogen and helium, along with trace amounts of other light gases. Like Jupiter, there isn't really a surface, and strong winds sweep clouds and storms along at speeds over 1,000 miles per hour. The rumors you may have heard about Saturn are true. It's less dense than water, which means that if you had a large enough swimming pool, Saturn would float on it like a giant beach ball!

Saturn's best known feature is its rings. When you gaze at the planet with the naked eye, most of the light comes from sunlight reflected off its rings. With a telescope, these rings look solid, but they are actually made up of trillions of ice-covered particles ranging in size from tiny dust grains to boulders the size of minivans, all orbiting the 74,000-mile-wide planet. The diameter of the ring system is about 200,000 miles, nearly the distance between the earth and our moon. Despite the width of the rings, they are actually thin, only a few miles thick at the most. Saturn's ring system is tilted by 27 degrees to its orbital path around the sun, so depending on where Saturn is in its orbit, we see it at different angles. At times, the rings are on edge from our view and thus pretty much disappear.

How did Saturn's rings come to be? More than thirty moons orbit Saturn, and astronomers believe that one of these moons smashed to bits from the big planet's tidal forces. The ring system developed out of the remnants of the smashed moon. This scenario is also used to explain the formation of the rings around the other giant gas planets.

Saturn's largest moon, Titan, interests astronomers because it has a thick nitrogen molecule atmosphere not unlike the earth, and it may have pools of liquid methane. Methane could prove to be a solvent that supports organic compounds and maybe life. The snag, though, is that the temperature on Titan is only 300°F below zero.

TELESCOPING SATURN

Saturn is just about the best thing to see through a telescope. You'll never forget the vivid ring system. See if you can spot a 1,000-mile-wide gap between the rings, called the Cassini gap. The earth's atmospheric turbulence affects how well you see Saturn, like anything else in the sky. If you don't see Saturn well one night, try the next.

You'll also see tiny "stars" that change position around Saturn from night to night. Those are some of Saturn's moons, available for your viewing pleasure.

URANUS AND NEPTUNE

Now we're getting out there. Uranus is 1.7 billion miles from the sun and Neptune is nearly 3 billion miles away. Like Jupiter and Saturn, Uranus and Neptune are gas giants, but they are not as big and are not made of the same stuff. Gaseous methane gives both planets their bluish-green color. Also, many moons endlessly circle both planets.

Let's clear up something here. The correct pronunciation of Uranus is yoor-a-nus, with a *short* "a." The other pronunciation leads to lots of jokes. Uranus used to be called George. Sir William Herschel, a musician-turned-astronomer, was hunting for new comets when he discovered the planet. In an

ABOVE, LEFT: Neptune (Courtesy of NASA/JPL-Caltech)

ABOVE, RIGHT: Uranus (Courtesy of NASA/JPL-Caltech)

effort to kiss up to the king of England, he named the planet George. Astronomers around the world were outraged by this human name when all the other planets were named after gods, so they renamed it after the ancient god Uranus. Personally, I like calling the planet George.

TELESCOPING URANUS AND NEPTUNE

It's possible to see Uranus with the naked eye, especially around opposition, but the skies have to be really dark. It's a sure bet with binoculars or telescopes, if you know where to look. I recommend the charts available from *Astronomy* or *Sky and Telescope* magazines, or Starry Night Enthusiast software. Even with lots of direction, though, Uranus is not all that rewarding. All you'll see is a blue-green dot.

PLUTO, NO LONGER A PLANET!

Since its discovery in 1930, Pluto was considered one of the solar system's nine planets, but in August 2006, it was officially defrocked as a planet by the International Astronomical Union. It's now considered a "dwarf planet," a new category of planets. Other dwarf planets include Ceres, the largest asteroid in the solar system, and the recently discovered Xena, a spherical icy world that's actually larger than Pluto.

There isn't much to say about Pluto because we really don't know much about it. It's over three billion miles away and takes 248 years to orbit the sun in a highly elliptical orbit that's inclined by over 17 degrees with respect to the eight planets.

Pluto is less than 1,500 miles in diameter, which is smaller than our moon. It is an icy little world with a thin nitrogen-ice atmosphere and three moons, including Charon that's also considered one of the new class of dwarf planets. Pluto, Charon, and Xena are also members of what could be billions of Kuiper objects that lie beyond the orbit of Neptune. Kuiper objects are icy, rocky objects, some of which eventually turn into comets.

Unless you have a huge telescope and surgically know your way around the night sky, forget about telescoping Pluto. It's just too far away! Even if you find it all you'll see is a faint dot.

Planet Data

Mercury

Diameter: 3,092 miles
Average Distance from Sun: 36 million miles
Rotation Period: 59 days
Sidereal Period: 88 days
Synodic Period: 116 days
Mass: .055 of Earth's
Surface Gravity: .38 of Earth's
Moons: 0

Venus

Diameter: 7,520 miles
Average Distance from Sun: 67.2 million miles
Rotation Period: 243 days (backwards)
Sidereal Period: 225 days
Synodic Period: 584 days
Mass: .82 of Earth's
Surface Gravity: .91 of Earth's
Moons: 0

Earth

Diameter: 7,936 miles
Average Distance from Sun: 93 million miles
Rotation Period: 24 hours
Sidereal Period: 365.25 days
Mass: 6 x 1,024 kilograms
Moons: 1

Mars

Diameter: 4,222 miles
Average Distance from Sun: 148 million miles
Rotation Period: 24.6 hours
Sidereal Period: 687 days
Synodic Period: 780 days
Mass: .11 of Earth's
Surface Gravity: .38 of Earth's
Moons: 2

Jupiter

Diameter: 88,000 miles
Average Distance from Sun: 484 million miles
Rotation Period: Almost 10 hours
Sidereal Period: 11.86 years
Synodic Period: 399 days
Mass: 318 times Earth's
Surface Gravity: 2.5 times Earth's
Moons: 58+

Saturn

Diameter: 75,000 miles
Average Distance from Sun: 888 million miles
Rotation Period: 10.7 hours
Sidereal Period: 29.5 years
Synodic Period: 378 days
Mass: 95 times Earth's
Surface Gravity: 1.14 times Earth's
Moons: 30+

Uranus

Diameter: 31,763 miles
Average Distance from Sun: 1.8 billion miles
Rotation Period: 17.2 hours
Sidereal Period: 84 years
Synodic Period: 369 days
Mass: 14.5 times Earth's
Surface Gravity: .90 of Earth's
Moons: 21

Neptune

Diameter: 30,775 miles
Average Distance from Sun: 2.8 billion miles
Rotation Period: 16.1 days
Sidereal Period: 164 years
Synodic Period: 367 days
Mass: 17.2 times Earth's
Surface Gravity: 1.14 times Earth's
Moons: 11

13

CELESTIAL EXTRAS

All sorts of celestial extras fill our night sky, including asteroids, comets, shooting stars, meteor showers, and aurora borealis. These objects and phenomena make for fun, ever-changing nights of stargazing. There are also many human additions taking a spin in our night sky, namely satellites, such as the International Space Station and the space shuttle. You have to work to see most of these celestial objects and phenomena, but that's what makes them fun!

Aurora borealis, also known as northern lights (Photograph © Jack Newton)

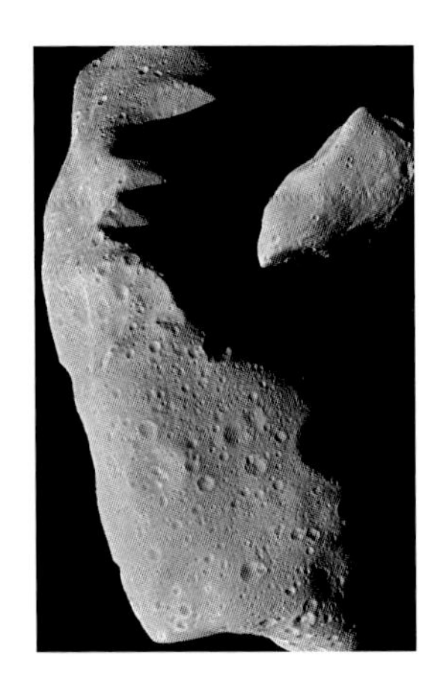

ABOVE: Asteroids Ida, left, and Gaspra, right (Courtesy of NASA/JPL-Caltech)

ASTEROIDS

Billions of years ago, after our solar system formed, small pieces of debris still zoomed around the heavens. Today we call these celestial bodies asteroids. These chunks of iron, nickel, and silicates come in all shapes and sizes, ranging in diameter from a few miles to 300 miles. To the naked eye, they look like tiny stars. You can also spot them with telescopes and keep track of their comings and goings with astronomy software or periodicals like *Astronomy* or *Sky and Telescope*.

Most asteroids orbit the sun in a belt between Jupiter and Mars, cleverly called the Asteroid Belt. Collisions in the Asteroid Belt send pieces of asteroids flying in all directions. The gravity of the giant planets, especially Jupiter, keeps these billions of asteroids in their place, though a few rebels exist elsewhere in the solar system.

The scary part about asteroids is that they are potential bullets for the earth. Several thousand asteroids have probably crossed the earth's orbit at one point or another. Can you say collision? Can you say

RIGHT: Comet Hale-Bopp in April 1997 (Photograph © Thomas Matheson)

doomsday? It's happened before and no doubt will happen again. It's theorized that, 65 million years ago, one such asteroid hit the Yucatan in Mexico and wiped out the dinosaurs. It's further theorized that a doomsday asteroid slams into the earth every 100 million years. An asteroid 10 miles in diameter would probably do-in civilization. It would create an explosion of a billion megatons! If the direct explosion didn't get you, so much dust would be kicked up that life on Earth would be choked out. Isn't that special?

COMETS

I like to describe comets as dirty cosmic snowballs. These celestial bodies basically consist of a boulder of ice, just 5 to 10 miles in diameter, followed by a spectacular vapor tail sometimes as long as 200 million miles. Comets mostly zoom about the Kuiper Belt and the Oort Cloud, both located past Uranus and Neptune in our outer solar system. Billions of comets orbit in the comet "bullpen" in a deep freeze. Many astronomers think much if not all of the earth's water came from comets.

Gravitational disturbances from passing stars or galactic tides send these snowballs hurling into elongated orbits within our inner solar system. Some end up in orbits that bring them back our way again and again. Examples are Halley's comet, which comes back toward the earth and the sun every seventy-six years, and Comet Temple-Tuttle, which comes back every thirty-three years. Some comets that swing into the inner solar system can be thrown back into the Kuiper Belt or Oort Cloud by the gravity of Jupiter, never to be seen again.

When a new comet comes in, it's exciting. Amateur astronomers discover more than 90 percent of new comets. Some folks dedicate themselves to hunting new comets. At first, they look like nothing more than fuzzy little stars. The payoff is that the new comet is named after the discoverer. Sometimes, two different people discover a comet simultaneously. Then the comet is given two names, such as comet Hale-Bopp, discovered in the 1990s.

When a comet gets close to the sun, it partially melts, creating a beautiful tail made up of two sections, the bluish ion tail and the dust tail, both pushed back by the sun's radiation pressure. These tails are about the closest thing to being nothing yet still being something. These vapor tails can stretch over 45 degrees in our sky.

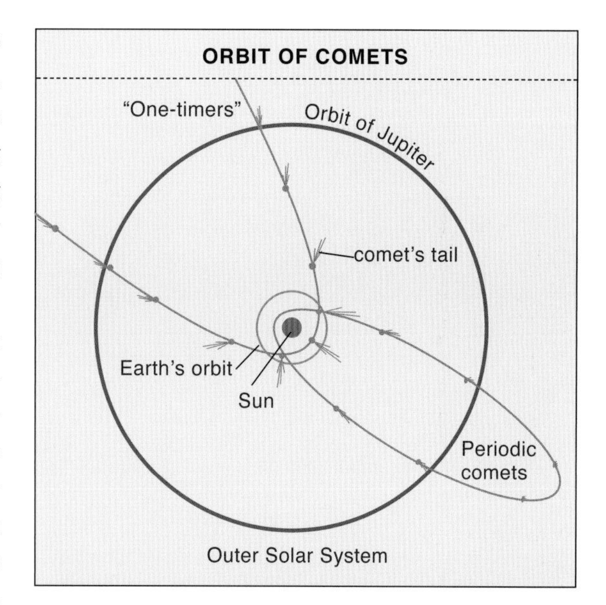

Comets throughout history have been objects of fear. Many cultures claimed they were harbingers of wars or other calamities. When Halley's comet came in 1910, the earth actually passed through part of its tail. Hucksters drummed up fears of poisonous gases in the comet's tail and got rich selling "anti-comet" pills.

Leonids meteor shower (Photograph © Thomas Matheson)

The only way a comet could harm the earth is if the nucleus plowed into us. A big-enough nucleus would cause great damage, even global catastrophe. During the summer of 1994, I watched through my 10-inch reflecting telescope as Jupiter got nailed by fragments from Comet Shoemaker-Levy. It was a rare celestial event that drew thousands of amateur astronomers to their scopes. The collisions left dark marks in Jupiter's clouds that lasted for weeks. An event like this only happens about once every thousand years. I feel lucky to have seen it!

SHOOTING STARS

Quick, what do you do when you see a shooting star? That's right, you make a wish. Of course, shooting stars aren't really stars at all. They are mostly natural space junk, usually asteroid debris that gets sucked into our atmosphere by the earth's gravity and burns up due to tremendous air friction.

MAJOR METEOR SHOWERS			
SHOWER	RADIANT	DATE OF MAXIMUM	AVERAGE NUMBER OF METEORS PER HOUR
Quadrantids	Draco	Jan. 3–4	50–100
Eta Aquarids	Aquarius	May 5	35
Delta Aquarids	Aquarius	July 29	25
Perseids	Perseus	Aug. 11–12	50–100
Orionids	Orion	Oct. 20–22	25
Leonids	Leo	Nov. 17	30–40
Geminids	Gemini	Dec. 13	50–100

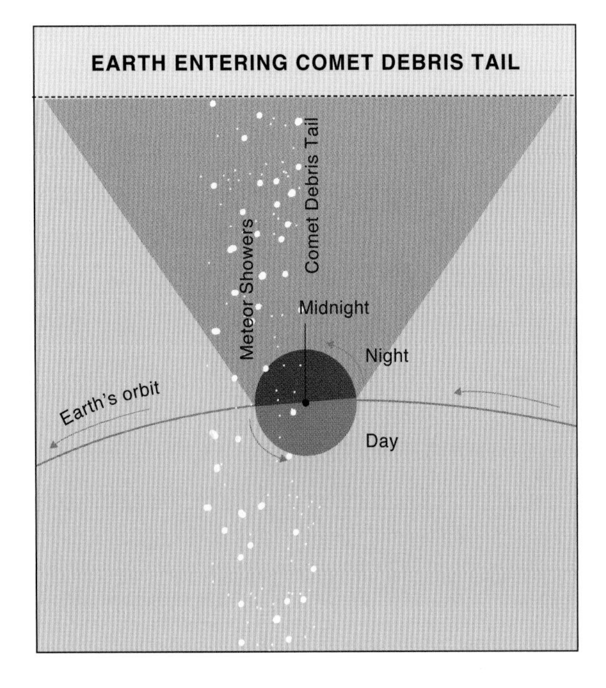

FACING PAGE: Aurora borealis (Photograph © Thomas Matheson)

They can hit the earth at over 40 miles a second. That's faster than a speeding bullet!

Shooting stars are called meteors, and if by chance they make it to the ground they're called meteorites. Most of them splash into the ocean and live with the fish. At times, they make it to terra firma, winding up in places like a cornfield in Nebraska or behind a Wal-Mart in Tallahassee, Florida. As far as I know, no one has been killed or seriously injured by a meteorite, but there are some wild stories.

The best story I've heard was how two houses, less than a mile apart, were hit by grapefruit-size meteorites in a span of eleven years. This happened in Wethersford, Connecticut. In both cases the meteorites went through the roof and landed on the floor. What are the odds of that?

Most meteors you see in the night sky are tiny, ranging in size from a grain of dust to a bit of gravel. As little as they are, though, they light up the sky. Combustion produces some of the light, but most of the light comes from ionization of the column of air through which the meteor passes. Electrons temporarily bounce away from the nuclei of atoms, producing bright light. When you see the tail of a faded meteor last for another second or two, you're witnessing that column of air atomically getting its act back together.

METEOR SHOWERS

Although asteroid debris produces most shooting stars, comets can also supply us with meteors—and meteor showers. Every year, as the earth orbits the sun, it passes into a debris trail left by a comet. When that happens we have a meteor shower. We get about twelve decent meteor showers every year, but the best are the Geminids in December and the Perseids in August. In the better meteor showers, you see well over fifty meteors an hour, barring moonlight.

Meteor showers are named after the constellation from which they appear to stream. That area is called the radiant. Don't just look at that particular part of the sky, which is a big mistake people make. Meteors in a meteor shower can be all over the sky. The radiant is just the general direction from which the meteors seem to diverge. The best way to watch a meteor shower is to get away from city lights, lie on the ground or in a reclining chair, and roll your eyes all around the sky. It's a lot of fun, especially with a big group of people. I've hosted many meteor shower parties and it's always a blast.

Meteor showers are best seen after midnight, especially just before morning twilight. That's when you're on the side of the earth that's heading into the comet debris trail. The principle is the same as driving on a highway on a summer night. You get a lot more bugs on your front windshield than you do on the rear window.

AURORA BOREALIS

Aurora borealis are the wild cards of the sky. These natural light shows have no real season. Some people think they are more common in winter, but that's only because the nights are longer. The truth is that aurora borealis, also called the northern lights, are not predictable.

Solar storms, what astronomers call coronal mass ejections, cause northern lights. Sunspots set

off these disturbances that send toward the earth massive, super-heated ionized-charged atoms, stripped of their electrons. The earth's magnetic field deflects most of these charged particles, but some follow the lines of the magnetic field and strike our atmosphere near the North or South Pole. When that happens we get a heck of a color show, as the charged particles react with atoms and molecules in our atmosphere. The aurora paints the sky all kinds of colors: green, white, red, blue, or all these colors mixed together. Most aurora borealis are seen around polar regions like Canada or Alaska, but they can spread down to our latitude and, sometimes, as far south as Florida. The South Pole counterpart of aurora borealis is aurora australis.

You might hear predictions of an aurora after astronomers observe a solar storm. Sometimes these predictions pan out; sometimes they don't. The best northern light shows I've seen were not predicted. What is the moral of the story? Any time you're out stargazing, keep an eye out for aurora.

Aurora borealis (Photograph © Bryan and Cherry Alexander)

SPACE JUNK

Since 1957, the dawn of the Space Age, over 26,000 satellites have been launched into orbit around the earth, and about a third of them are still out there. Some satellites, like the International Space Station, even have people in them. A lot of junk flies around up there: useless, out of control satellites; spent rocket stages; and other random space garbage. It's like a floating junkyard.

Just about any time you stargaze, you'll see star-like objects march across the heavens, either a little after evening twilight or just before morning twilight. You're almost guaranteed to see at least one satellite zipping among the constellations. Satellites are seen mostly at those times because they're better able to reflect sunlight then. That's the only light they have! You could shine the biggest spotlight in the world from a satellite and we'd never see it. These satellites are at least 110 miles high in the sky, but most of the healthy ones are even higher.

Most military reconnaissance and communications satellites are high flyers, cruising along at altitudes of 200 to over 1,000 miles. They traverse our sky from horizon to horizon in two to three minutes. Faster-paced satellites take much lower orbits and, in many cases, are not long for the orbital world. The drag of the upper atmosphere slowly pulls down these lower-trajectory satellites, leaving them more and more vulnerable to being totally sucked in by the atmosphere and incinerated by air friction.

Most satellites travel west to east across the sky, taking advantage of the earth's east-to-west rotation, which is faster toward the equator. That's why just about all U.S. space launches take place at Cape Canaveral, Florida, as opposed to Fairbanks, Alaska. Some satellites travel in polar orbits, south to north or vice versa. A lot of those are spy satellites, so wave when you see them! Some satellites are geostationary, sitting in one place above the earth, their orbits in perfect sync with earth's rotation.

The brightest satellites are the International Space Station and the space shuttle. They shine nearly as bright as highflying aircraft, sometimes more so when they're docked together. Another fun group of satellites to watch are the sixty-plus Iridium satellites that provide worldwide cell-phone service. As they track across our heavens, they dramatically flare up for a few seconds, sometimes over 100 times brighter than the brightest stars. This temporary flash is caused by their highly reflective solar panels kicking reflected sunlight our way.

International Space Station (Photograph © Comstock)

You can use Websites and software such as the Starry Night Enthusiast program to keep track of every major spacecraft and satellite in your sky. At the very least, look for the space shuttle when it heads over this area. If the sun's angle is just right, you can sometimes see a faint patchy cloud ahead of the shuttle. You know what that is? It is toilet and other waste dumped by the shuttle. It precedes the shuttle in the sky because of the law of conservation of angular momentum. When the astronauts dump waste, they shoot it out and below the shuttle so it can't get sucked back in. They don't want a rebound!

With the potty cloud winding ahead of the space shuttle, it truly becomes number one and two on the runway of space!

14

YOU AND YOUR TELESCOPE

So you want to buy a telescope. What do you buy and where do you buy it? Let me answer the second question first. Buy a telescope from a place where people know about telescopes. Ask lots of questions. There's no such thing as a stupid question. This is a big investment. Also, be wary of a buying a scope on the Internet. Unless you know what you're doing, you can get taken to the celestial cleaners.

Sagittarius Cluster, M22 (Photograph © Mark Hansen)

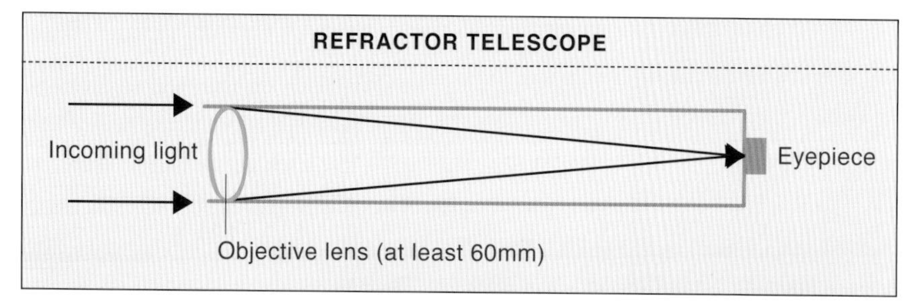

ABOVE: Refractor telescope (Photograph © Denny Long)

RIGHT: Reflector telescope (Photograph © Denny Long)

When you're buying a telescope, remember this: the most important aspect of a scope is light-gathering ability. The more light you gather into your telescope, the clearer the image will be. And isn't that the most important thing?

You can put the right eyepiece in just about any telescope to crank it to 300-power magnification, but if the image is fuzzy, what good is it? Most of the time, you'll be perfectly happy with 100- to 200-power magnification. Generally, the wider the telescope, the more light it lets in. When it come to telescopes, fat is good!

When you start using your telescope, you may be disappointed that you're not seeing some of the vivid colors found in the Hubble Space Telescope's photographs. The truth is that a lot of these pictures are colorized for both scientific and showbiz purposes. What you see through your telescope may not be as colorful, but it is generally more realistic.

So what kind of telescope should you get? There are three main choices: refractor, reflector, or Schmidt-Cassegrain.

REFRACTOR TELESCOPE

When most people think of a telescope, a picture of a refractor telescope pops in their brain. It's a simple design. Light is gathered into the telescope by the objective lens at one end. The light refracts or bends inside the tube, and the object is viewed through an eyepiece at the other end of the telescope.

Telescopes are sold by the width of the objective lens. A 60-mm refractor is the minimum size to buy. Anything less than that and you're getting a toy. The telescope should come with several eyepieces that will give you different magnifications. You'll pay $150 to $200 for a decent 60-mm refractor. A 90-mm refractor will do a nice job, but of course you'll pay more. No matter which telescope you buy, make sure that the mount is good and sturdy. Nothing is worse than trying to use a telescope on a flimsy mount.

REFLECTOR TELESCOPE

Reflector telescopes are my favorite, and I daresay they're the favorite of most amateur astronomers. Sir Isaac Newton invented this kind of telescope, which is more properly called a "Newtonian Reflector." I think you get more celestial bang for your buck with a reflector telescope. The drawback is that they can get big and bulky, though most can be broken down into smaller pieces for transportation and then easily reassembled. For example, my large 14.5-inch Dobsonian reflector can be broken down to fit in my wife's Geo Metro.

Reflector telescopes gather light with a concave parabolic mirror at the back of the tube. The wider the mirror, the more light you gather and the clearer your image. The mirror reflects incoming light to the focal point at the front of the telescope tube. A secondary flat mirror at a 45-degree angle takes the image through the side of the scope and through a mounted eyepiece. You aim the telescope with a small finder telescope on the side of the tube.

Reflector telescopes are sold by their mirror diameters. The smallest reflectors have mirrors around 3 inches wide, and you can generally buy them for less than $100. They are great for kids because they can be easily held and aimed. The next step up is a 4.25- or 4.5-inch reflector for about $200.

If you can afford it, I recommend a 6-inch reflector. You can pick one up for less than $400, and I think you'll be happy with it for a long time. Of course, you can buy an 8-, 10-, 16-, or even a 30-inch reflector, but you'll pay a lot more and have a monster on your hands.

For mounting, you have a choice between an equatorial mount that will help you follow a celestial object across the sky, or a less expensive Dobsonian mount that is generally easier to use.

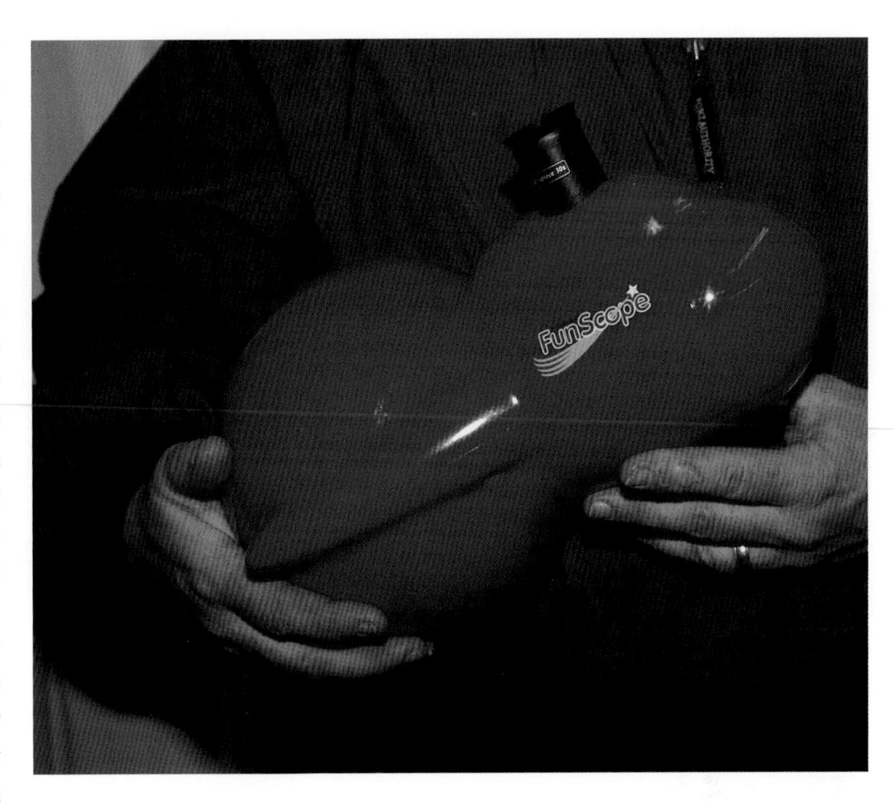

ABOVE: Reflector "starter" telescope (Photograph © Denny Long)

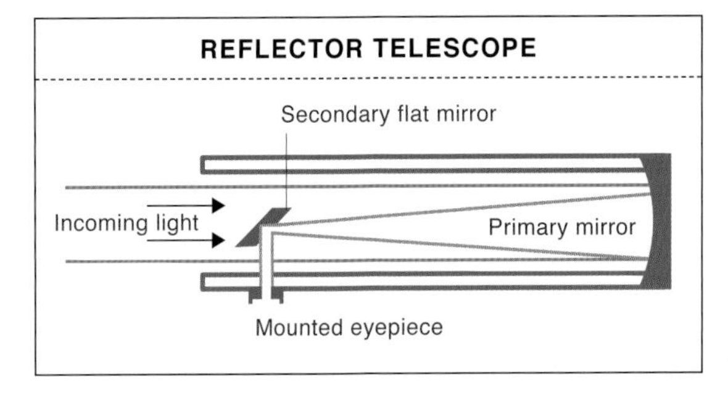

SCHMIDT-CASSEGRAIN TELESCOPE

These scopes are spinoffs of reflectors. Light is gathered by a mirror and reflected to the other side of the telescope, where a special mirror redirects the reflection back through a hole in the mirror and then back to a mounted eyepiece. This makes the Schmidt-Cassegrain shorter and a lot more portable but also a lot more expensive. The images are not as bright as a reflector telescope because the incoming light has to go through more glass.

The minimum aperture I would get for this scope is 5 or 6 inches. Some Schmidt-Cassegrain scopes are heavily marketed but have only 90-mm and 125-mm apertures. I don't think they are worth the money. With the 90-mm scope, you can hardly see Saturn's rings.

COMPUTERS AND ELECTRONICS

Computers, software, and other electronics are becoming more popular for use with telescopes and for good reason: they can save you a lot of time and help you find things. Many telescopes have motor drives that actually follow celestial objects across the sky so they won't drift out of your eyepiece. This feature can be a real lifesaver.

Some telescopes come with GPS-type computers that help you find what you're looking for. The computer may have thousands of celestial objects

Astrophotography

FACING PAGE: Star trails (Photograph © Gary Braasch)

If you're new to amateur astronomy—and I have a feeling a lot of you are—you may want to get used to viewing the night sky with a telescope before you try astrophotography. But if you're up to the challenge, here are some methods.

STAR TRAILS

This form of astrophotography is the easiest but you do need a single lens reflex (SLR) camera attached to a tripod. I recommend a shutter cable switch so you can open the shutter without disturbing the camera. Disable the flash unit and use color print film with an ISO speed of around 200. Set your camera for long-term exposure. For most cameras, this is done by setting it on "B" or bulb. Set the focus to infinity and the aperture to the widest setting, like an f/1.8.

When your camera is ready, point it at the sky and open the shutter for fifteen to thirty seconds. If you keep the shutter open too long, stray ambient light will overexpose your picture. The darker your sky, the longer you can keep the shutter open.

When you get your pictures back, the stars show up as curved lines, reflecting their movement in the sky during the time the shutter was open. The photographs may show more stars and colors than you see. It's especially fun to point your camera at Polaris for a circular star trail photograph. You may not get great pictures the first time, but keep at it, varying the exposure time.

CAMERA-TELESCOPE MATING

You can mount your camera on the side of your telescope or attach it to the eyepiece mount on your telescope. In both cases, you will need a clock drive system to follow the stars across the sky or you'll get star-trail pictures.

If you want to mount the camera on the side of your telescope, contact a photo shop or a telescope supply center for specialized mounting devices. You'll need another kind of adapter if you mount your camera to your eyepiece mount.

DIGITAL PHOTOGRAPHY

Digital photography is the biggest advancement in astrophotography, but it can get expensive, especially if you use a charged coupled device (CCD) camera. Instead of gathering the image on film, it uses a silicon chip that is more efficient for gathering light. This is how a video camera gathers light. The downside here is that CCD cameras and the corresponding software cost well over $1,000. There's also a trial-and-error period before you get the desired images.

WEBCAMS AND YOUR TELESCOPE

Using a webcam is the latest trend in astrophotography, especially to get great pictures of the planets. Webcams are a heck of a lot cheaper than CCD cameras, but you do need to attach your telescope to a laptop or desktop computer, and you need software to process the pictures.

If you want to get into heavy-duty astrophotography, you'll find great articles and advertisements in *Astronomy* and *Sky and Telescope*. Astrophotography advances all the time, and these magazines offer the best sources to find out about the latest developments.

Schmidt-Cassegrain telescope (Photograph © Denny Long)

in its library. You punch in what you want and the computer helps you find it with a digital readout. Some telescopes even have "go-to" computers and motors that actually point the telescope at whatever celestial object you choose in the computer's vast library.

These electronic devices and computers are great. Just make sure your telescope optics are sound. That's the meat on the table. All the rest are just the trimmings.

TELESCOPE TECHNIQUES

The most important technique to remember when you're using any telescope, especially in cooler weather, is to set your telescope up and let it sit out at least thirty to forty-five minutes before you use it. Let your eyepieces sit outside just as long. This is the number-one rule! Your telescope has to acclimate to the outside temperature, otherwise your images blur.

Also, never try to observe from the inside of your home. The glass panes in your window will mess up the image. Opening a window or door and sticking the front of your scope outside isn't any good either, because currents of warm air from your house will interfere with your image.

Set your telescope outside on firm ground. Avoid setting up your scope on a wooden deck. No matter how well it's built, vibrations from your movements will jiggle the scope just enough to drive you crazy. The best place to set up your scope is a flat, grassy area. A driveway or a parking lot is not as good because heat waves, gathered earlier in the day, may still be rising from the surface.

When you're setting up, make sure the small finder telescope mounted on the side of your telescope is properly aligned. This is important because if it's out of alignment, finding any celestial object in the night sky will be frustrating to impossible. Finder scopes get bumped around and should be readjusted every time you use your telescope. Your finder scope is low power and should have crosshairs. The best way to adjust it is to put a low-power eyepiece in your main telescope and then aim your finder at some prominent object along the horizon, like a church steeple or a flag. Center it right at the intersection of the crosshairs. Then see if the object is in the center of your viewing field in your main telescope. If it is, you're good to go, but if it's not, move your main scope slowly around until that object is centered in your big scope. Once it is, see where your test target is in your finder scope and then adjust it so the target is right on the crosshairs. You should be able to adjust your finder scope by turning hand screws on the mounting. Check out the instructions that came with scope so you don't break anything.

With whatever you observe through your telescope, start out with a high-focal-length/low-power eyepiece. The focal length is usually labeled on the eyepiece. A 25- to 40-mm focal-length eyepiece should do fine. Once you get your target in view, increase the magnification gradually with lower-focal-length/higher-power eyepieces. The image will become slightly fuzzier with each increase in magnification, depending on the size of your telescope. To figure out how much magnification you have, divide the focal length of your telescope by the focal length of your eyepiece. Your telescope's focal length should be listed in the specifications or on the telescope itself. For example, if the focal length of your telescope is 1500 mm and you're using a 15-mm eyepiece, you have 100-power magnification, which means you're seeing your target 100 times larger than you would with your naked eye.

Look for objects at least 25 to 30 degrees above the horizon, or a third of the way up from the horizon to the overhead zenith. If you aim your scope at something closer to the horizon, you look through more of the earth's atmosphere and blur your image. On some nights, even if you look above 30 degrees, your object may not be impressive because of atmospheric turbulence caused by high winds or the jet stream. Some nights just have what amateur astronomers call "bad seeing." It may seem like a clear night, but the sky is not all that wonderful through the telescope. Don't be discouraged. Just try looking another night. With whatever you're viewing, especially the planets, try to look continually for at least ten to fifteen minutes so your eye will get used to the darkness inside the eyepiece field. The longer you look, the more detail you'll see. Quick glances don't cut it with serious telescoping!

A good pair of binoculars can be a telescope's best friend. Wear a light pair around your neck for quick scouting. The best binoculars are 7 x 35. The magnification power is 7 and the aperture in millimeters is 35. That's all you need optically.

When you're looking for a Messier or NGC object, plot out your strategy for finding it with the star maps and constellation charts. This method is called star hopping. Find a moderate bright star that's near your target object and star hop from that star to your target. Your finder scope will be instrumental. You may have to hop from several stars to get your target in the eyepiece. Again, use a low-power eyepiece, because it has a wider field of view for finding those celestial deep tracks.

SCHMIDT-CASSEGRAIN TELESCOPE

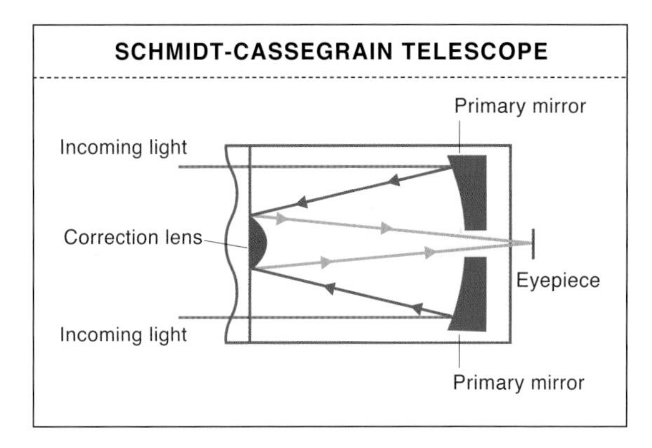

Above all, be patient with your telescope. Read the instructions. You don't have to conquer the whole universe in one night or even one year. Amateur astronomy is a lifelong hobby and for many, including myself, a passion. You may want to contact or join a club. Share your knowledge with other people as passionate as you. You'll learn from them and they'll learn from you!

Appendix A

Planet Locator

Saturn (Photograph © Robert Gendler)

Use this guide to locate the four brightest planets in our sky through 2017. For more detailed planet wanderings, consult *Astronomy* or *Sky and Telescope* magazines, use computer software such as Starry Night Enthusiast, or check out my website, www.lynchandthestars.com.

Venus

The third-brightest object in our sky after the sun and the moon, Venus oscillates between being a "morning star" in the east and an "evening star" in the west. An inferior planet, Venus never strays more than 48 degrees from the sun.

February 2007 to early August 2007	Evening
Early September 2007 to March 2008	Morning
Early September 2008 to mid-March 2009	Evening
Mid-April 2009 to mid-November 2009	Morning
Late March 2010 to mid-October 2010	Evening
Early November 2010 to mid-June 2011	Morning
November 2011 to mid-May 2012	Evening
Late June 2012 to December 2012	Morning
Mid-June 2013 to December 2013	Evening
February 2014 to mid-August 2014	Morning
Late January 2015 to late July 2015	Evening
Late August 2015 to late March 2016	Morning
Early August 2016 to early March 2017	Evening
Mid-April 2017 to late November 2017	Morning

Look for the superior planets Mars, Jupiter, and Saturn about a month before opposition, when they rise in the east one to two hours after evening twilight. These planets linger in our night sky two to three months after opposition, when they are visible in the west just after evening twilight.

Mars

DATE OF OPPOSITION	MINIMUM DISTANCE FROM EARTH	BEST VIEWING
12/24/2007	54.8 million miles	December 2007 to March 2008
01/29/2010	61.3 million miles	January 2010 to April 2010
03/03/2012	62.3 million miles	February 2012 to May 2012
04/08/2014	58 million miles	March 2014 to June 2014
05/22/2016	47.3 million miles	April 2016 to July 2016

Jupiter

DATE OF OPPOSITION	MINIMUM DISTANCE FROM EARTH	BEST VIEWING
06/06/2007	400 million miles	May 2007 to August 2007
07/09/2008	386 million miles	June 2008 to September 2008
08/14/2009	374 million miles	July 2009 to October 2009
11/21/2010	371 million miles	August 2010 to November 2010
10/27/2011	369 million miles	September 2011 to December 2011
12/01/2012	378 million miles	October 2012 to January 2013
01/04/2014	392 million miles	December 2013 to March 2014
02/04/2015	404 million miles	December 2014 to April 2015
03/08/2016	412 million miles	January 2016 to May 2016
04/07/2017	419 million miles	March 2017 to June 2017

Saturn

DATE OF OPPOSITION	MINIMUM DISTANCE FROM EARTH	BEST VIEWING
02/19/2007	764 million miles	Mid-January 2007 to May 2007
02/23/2008	771 million miles	February 2008 to early June 2008
03/08/2009*	781 million miles	February 2009 to June 2009
03/21/2010*	791 million miles	March 2010 to June 2010
04/03/2011*	801 million miles	March 2011 to early July 2011
04/15/2012	811 million miles	Mid-March 2012 to mid-July 2012
04/27/2013	820 million miles	Late March 2013 to late July 2013
05/09/2014	828 million miles	Early April 2014 to July 2014
05/23/2015	837 million miles	Late April 2015 to August 2015
06/03/2016	837 million miles	May 2016 to August 2016
06/15/2017	840 million miles	May 2017 to August 2017

* Saturn's rings appear nearly edge-on.

Appendix B

Brightest Stars in Southern California

The red giant Antares is the tenth-brightest star in Southern California skies. It is so bright it can obscure the Cat's Eye Cluster, M4, in the constellation Scorpius. (Photograph © Rick Krejci)

STAR	MEANING	MAGNITUDE	CONSTELLATION	LUMINOSITY (TIMES SUN'S)	DISTANCE (LIGHT YEARS)	TEMPERATURE (FAHRENHEIT)	ESTIMATED DIAMETER (MILES)
Sirius	The scorcher	-1.46	Canis Major	27	8.6	15,485	1.9 million
Canopus	The helmsman	-.63	Carina	21,000	313	14,000	65 million
Arcturus	Bear keeper	-.04	Boötes	298	36.7	6,931	29.4 million
Vega	Diving eagle	.03	Lyra	61	25.3	15,583	2.8 million
Capella	She-goat	.08	Auriga	162	42.2	9,294	12.1 million
Rigel	The foot	.12	Orion	51,194	777	15,877	74.9 million
Procyon	Prelude to the dog	.38	Canis Minor	7.5	11.4	11,780	1.7 million
Betelgeuse	Armpit of the great one	.5	Orion	58,980	429	5,819	1 billion
Altair	The flyer	.77	Aquila	12	16.8	13,504	1.6 million
Aldebaran	Follower	.85	Taurus	1,080	65	5,671	82 million
Antares	Rival of Mars	.96	Scorpius	387,000	604	4,537	Unknown
Spica	Ear of wheat	.98	Virgo	14,780	263	36,204	8.4 million
Pollux	A lot of wine	1.14	Gemini	49	34	8,170	8.6 million
Fomalhaut	Fish mouth	1.16	Piscis Australis	19	25	14,185	1.6 million
Deneb	Tail of the hen	1.25	Cygnus	301,059	3,262	14,678	224 million
Regulus	The prince	1.35	Leo	221	78	18,885	4 million
Adhara	Young lady	1.5	Canis Major	18,042	432	32,015	12.7 million
Castor	Centaur	1.58	Gemini	59	52	15,236	2.9 million
Shaula	Raised tail	1.63	Scorpius	57,140	709	34,446	17.2 million
Bellatrix	Lady warrior	1.64	Orion	6,000	243	34,181	6 million
Elnath	Sea monster	1.65	Taurus	669	131	22,330	4.7 million
Alnilam	Pearl arrangement	1.7	Orion	112,435	1,360	28,165	38 million
Alnitak	The belt	1.77	Orion	47,500	825	30,209	22 million
Alioth	Goat	1.77	Ursa Major	127	81	15,796	4 million
Dubhe	Bear	1.79	Ursa Major	416	124	7,817	27.6 million

Appendix C

Resources

PLANETARIUMS & OBSERVATORIES

Bakersfield College Planetarium. www.bc.cc.ca.us/planetarium.

Donald E. Bianchi Planetarium, California State University-Northridge. www.csun.edu/phys.

El Camino College Planetarium, Torrance. www.elcamino.edu/academics/naturalsciences/astronomy.

George F. Beattie Planetarium, San Bernardino Valley College. www.valleycollege.edu/Facilities/Planetarium.

Gladwin Planetarium, Santa Barbara Museum of Natural History. www.sbnature.org.

Glendale Community College Planetarium. www.glendale.cc.ca.us/planetarium.

Griffith Observatory, Los Angeles. www.griffithobs.org.

John Drescher Planetarium, Santa Monica College. http://events.smc.edu/planetarium.

Lick Observatory, University of California-Santa Cruz, Mount Hamilton. http://mthamilton.ucolick.org.

Millikan Planetarium and Frank P. Brackett Observatory, Pomona College, Claremont. www.astronomy.pomona.edu.

Mount Laguna Observatory, San Diego State University. http://mintaka.sdsu.edu.

Mount Wilson Observatory, Pasadena. www.mtwilson.edu.

Palomar College Planetarium, San Marcos. www.palomar.edu/planetarium.

Palomar Observatory, California Institute of Technology, Palomar Mountain. www.astro.caltech.edu/palomar.

Reuben H. Fleet Science Center Planetarium, San Diego. www.rhfleet.org/astronomy.html.

Robert Brownlee Observatory, Lake Arrowhead Village. www.mountain-skies.org/RBO.html.

Sam B. Peña Planetarium, Impact Center, Visalia. www.tcoe.k12.ca.us/ImpactCenter/Planetarium.shtm.

Stony Ridge Observatory, Temecula. www.stony-ridge.org.

Tessmann Planetarium, Santa Ana College. www.sac.edu.

University of California-Los Angeles Planetarium. www.astro.ucla.edu/planetarium.

CLUBS & SOCIETIES

Antelope Valley Astronomy Club, Lancaster. www.avastronomy-club.org.

Astronomical Society of the Desert, Rancho Mirage. www.astrorx.org.

Central Coast Astronomical Society, San Luis Obispo. www.ccastronomy.org.

China Lake Astronomical Society, Ridgecrest. www1.iwvisp.com/brower/clas.html.

Colorado River Astronomy Club, Blythe. www.home.earthlink.net/~astroclub.

FirstLight Astronomy Club, Temecula. www.firstlightastro.com.

High Desert Astronomical Society, Apple Valley. www.hidasonline.com.

Local Group-Astronomy Club of Santa Clarita Valley. www.lgscv.org.

Los Angeles Astronomical Society. www.laas.org.

Los Angeles Valley College Astronomy Club. www.astrophys-assist.com/lavc.

Mountain Skies Astronomical Society, Lake Arrowhead. www.mountain-skies.org.

Oceanside Photo and Telescope Astronomical Society. www.astronomyoutreach.net/optas.

Orange County Astronomers, Costa Mesa. www.ocastronomers.org.

Polaris Observatory Association, Lockwood Valley. www.frazmtn.com/~polaris.

Pomona Valley Amateur Astronomers, Upland. www.pvaa.us.

Riverside Astronomical Society. www.rivastro.org.

San Bernardino Valley Amateur Astronomers, Redlands. www.sbvaa.org.

San Diego Astronomy Association. www.sdaa.org.

Santa Barbara Astronomical Unit. www.sbau.org.

Santa Cruz Astronomy Club. www.astro.santa-cruz.ca.us.

Sidewalk Astronomers, Hollywood. www.sidewalkastronomers.us.

South Bay Astronomical Society, Redondo Beach. www.geocities.com/sbas_elcamino.

Vandenberg Amateur Astronomical Society, Vandenberg Air Force Base. www.members.aol.com/vaastronomy.

Ventura County Astronomical Society, Simi Valley. www.vcas.org.

SOFTWARE

Starry Night Enthusiast. Get this! It is the best software I've seen. Easy to use and full of the latest data, this software allows you to travel to the planets and make just about any star map. For more advanced stargazing software, try Starry Night Pro. Both programs are available at computer software stores and www.starrynight.com.

MAGAZINES

Astronomy. Great for beginners. www.astronomy.com.

Sky and Telescope. Wonderful, but aimed at more advanced stargazers. www.skyandtelescope.com.

WEBSITES

Astronomy Picture of the Day. Great site. http://antwrp.gsfc.nasa.gov/apod/astropix.html.

Constellations, The. www.dibonsmith.com/menu.htm.

Heavens Above. This site tracks space junk, including the space shuttle, the International Space Station, iridium flares, and other satellites. www.heavens-above.com.

Jack Horkheimer: Star Gazer. Wonderful site! www.starhustler.com.

Live Sky. www.livesky.com

NASA. www.nasa.gov.

Space.com. Great site for news and education. www.space.com.

Space Weather. www.spaceweather.com

StarDate. Features streaming audio! www.stardate.org.

FACING PAGE: A reflection nebula, NCG1999, in the constellation Orion. (Courtesy NASA, the Hubble Heritage Team)

Index

Appendix D
Monthly Star Maps

The constellation Cassiopeia shines brightly in the summer sky. (Photograph © Thomas Matheson)

JANUARY

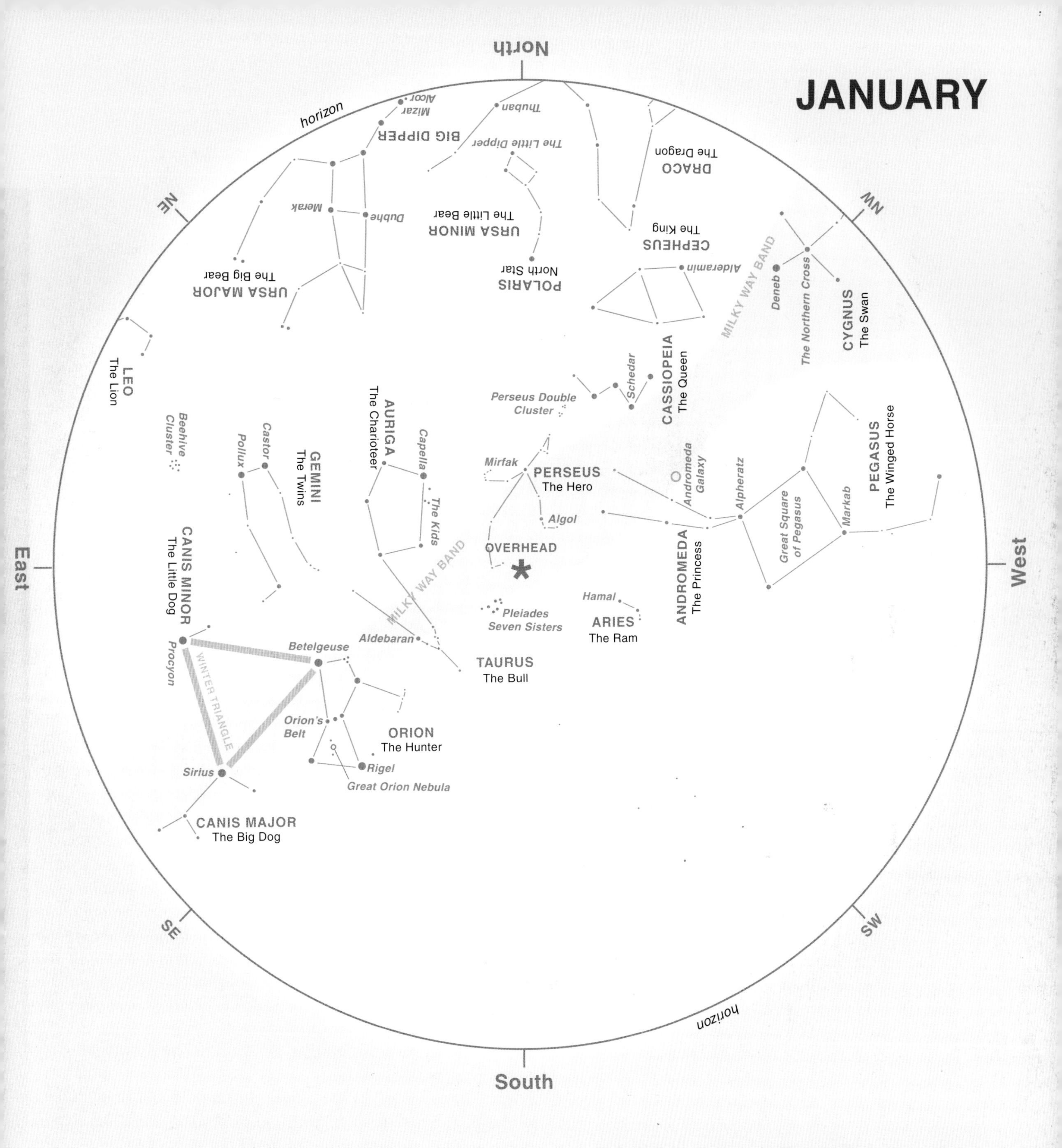

Use this map at the following times:

8 P.M. IN JANUARY
MIDNIGHT IN NOVEMBER
4 A.M. IN SEPTEMBER

FEBRUARY

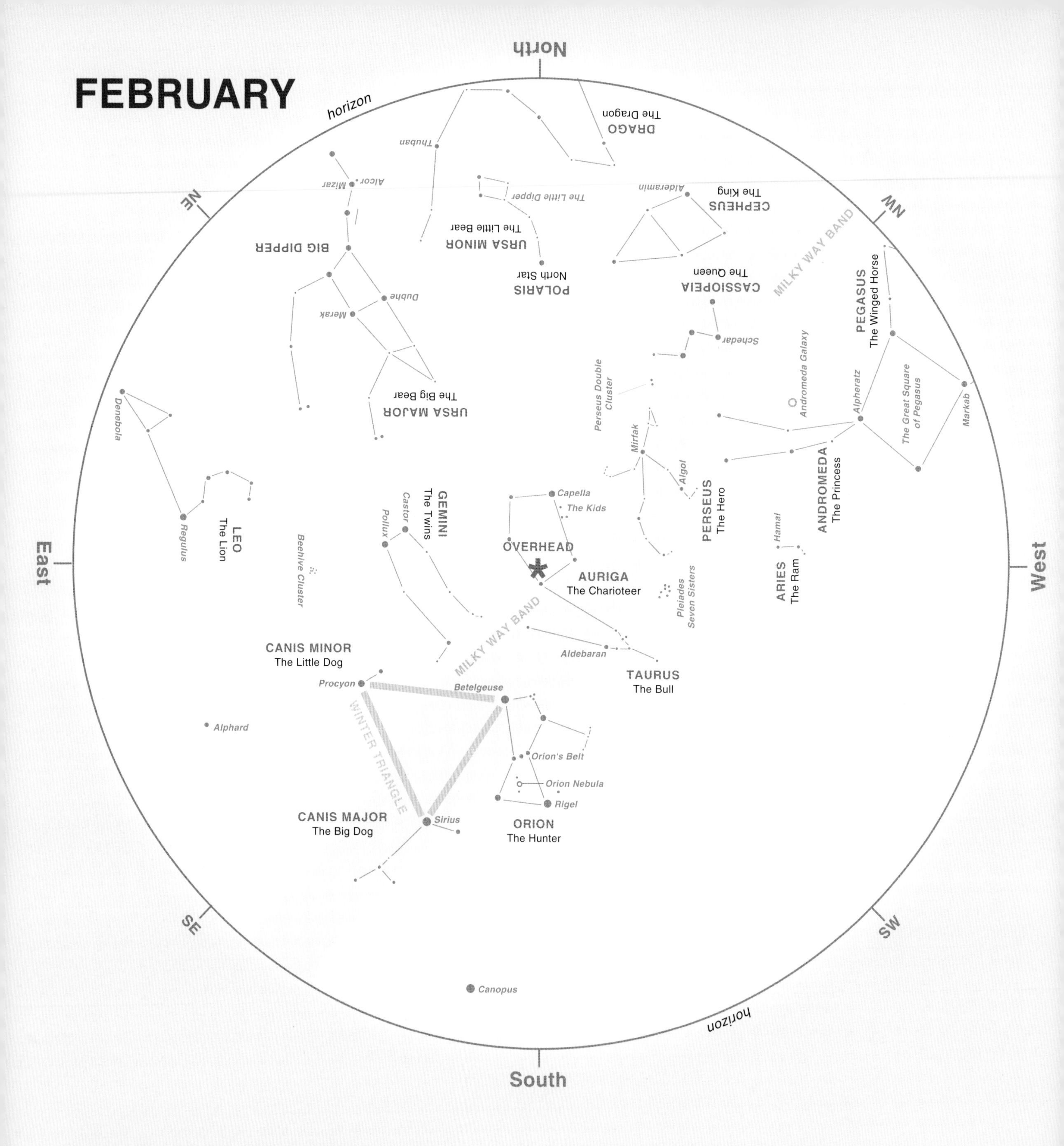

Use this map at the following times:

8 P.M. IN FEBRUARY
MIDNIGHT IN DECEMBER
5 A.M. IN OCTOBER

MARCH

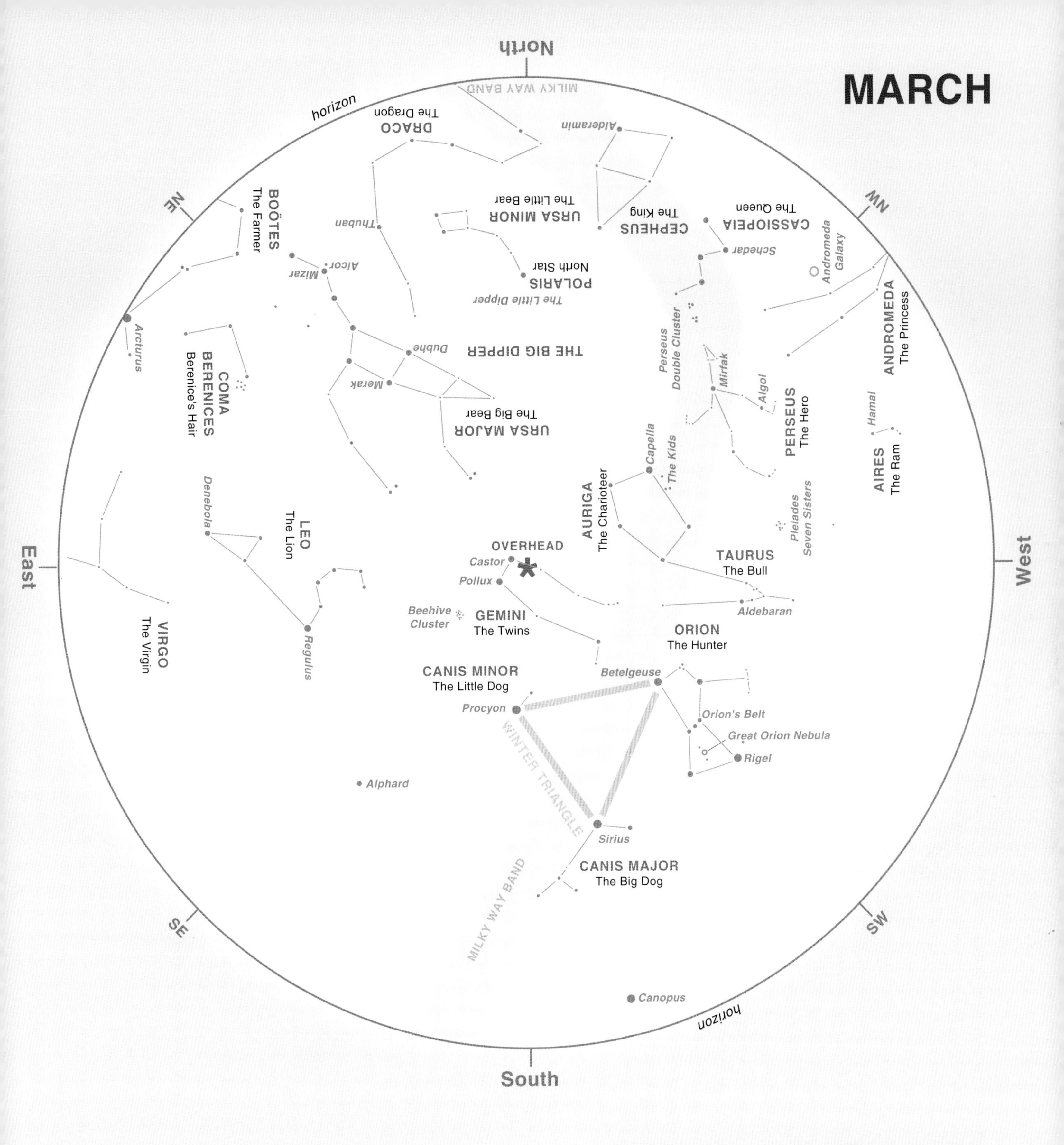

Use this map at the following times:

8 P.M. IN MARCH
MIDNIGHT IN JANUARY
5 A.M. IN NOVEMBER

APRIL

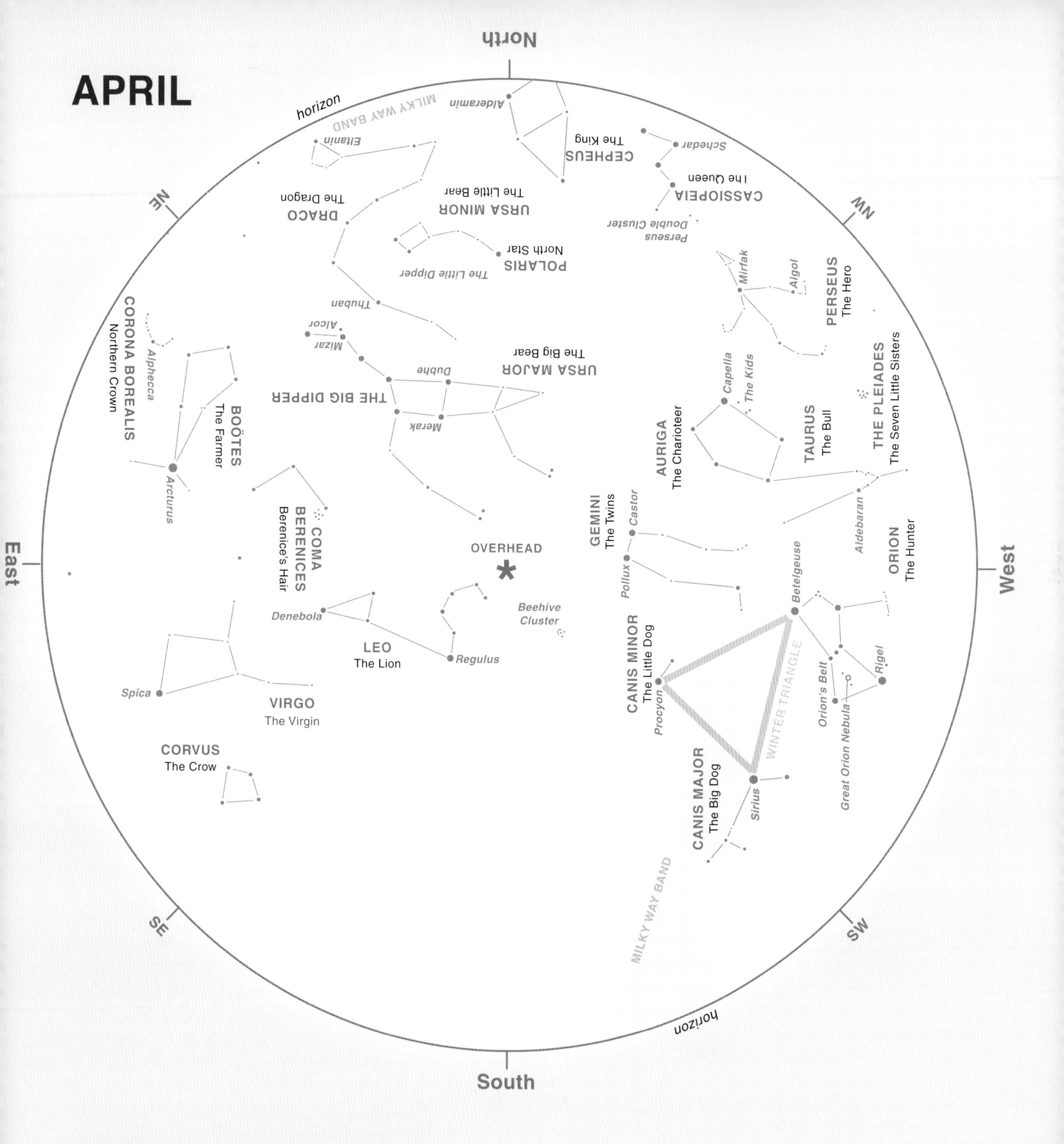

Use this map at the following times:

9 P.M. IN APRIL
1 A.M. IN FEBRUARY
5 A.M. IN DECEMBER

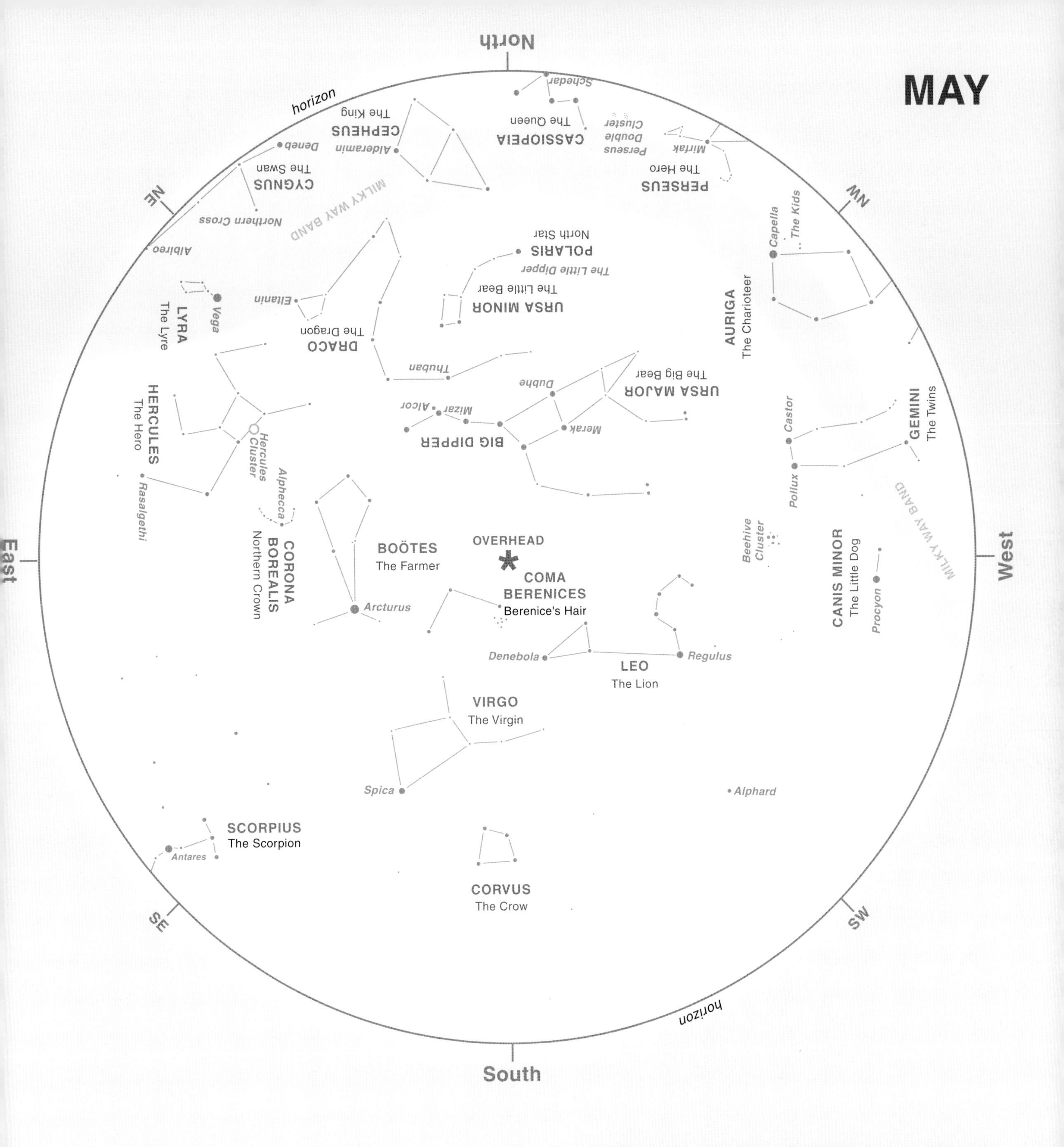

Use this map at the following times:

10 P.M. IN MAY
1 A.M. IN MARCH
6 A.M. IN JANUARY

JUNE

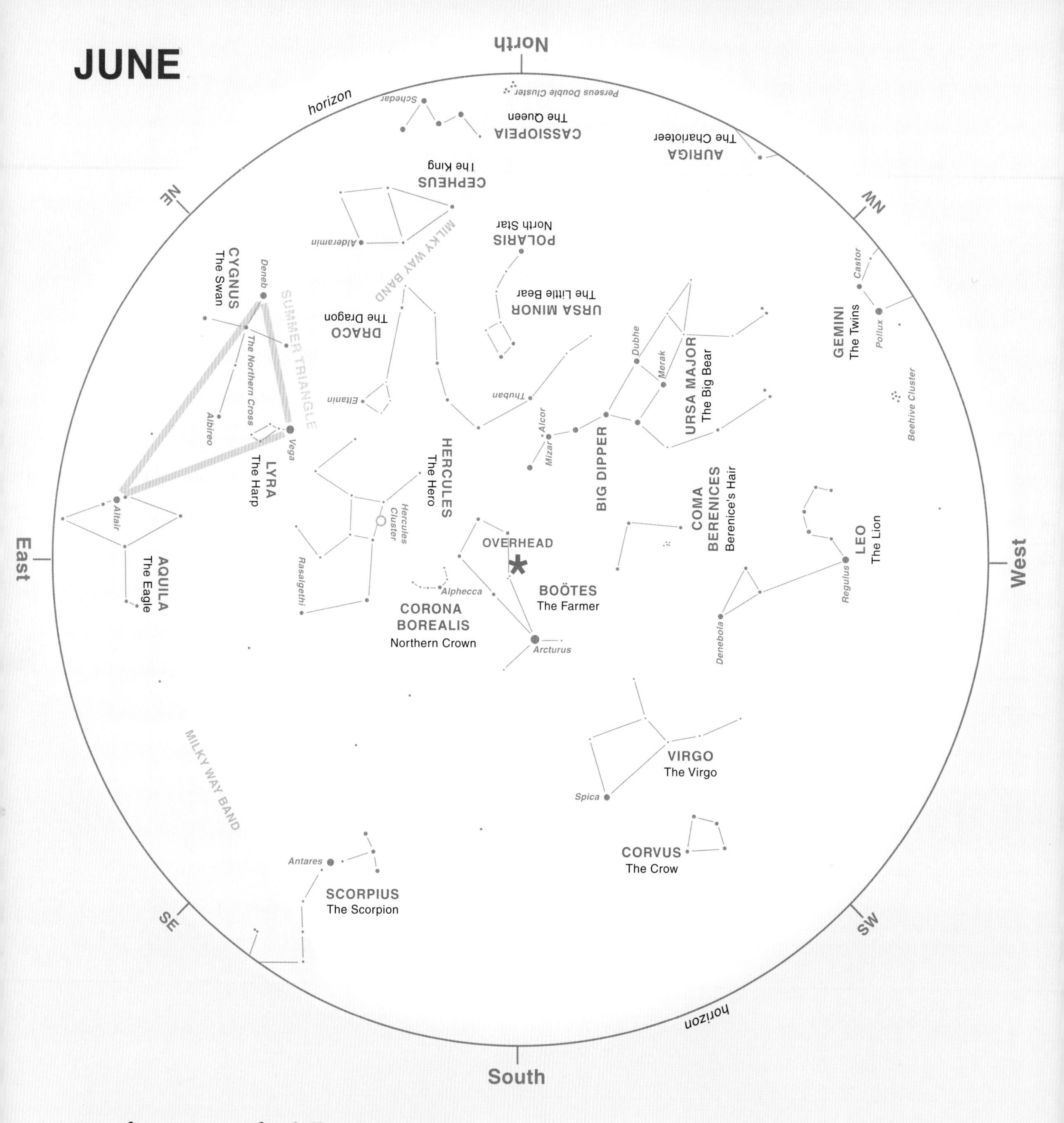

Use this map at the following times:

10 P.M. IN JUNE
2 A.M. IN APRIL
6 A.M. IN FEBRUARY

JULY

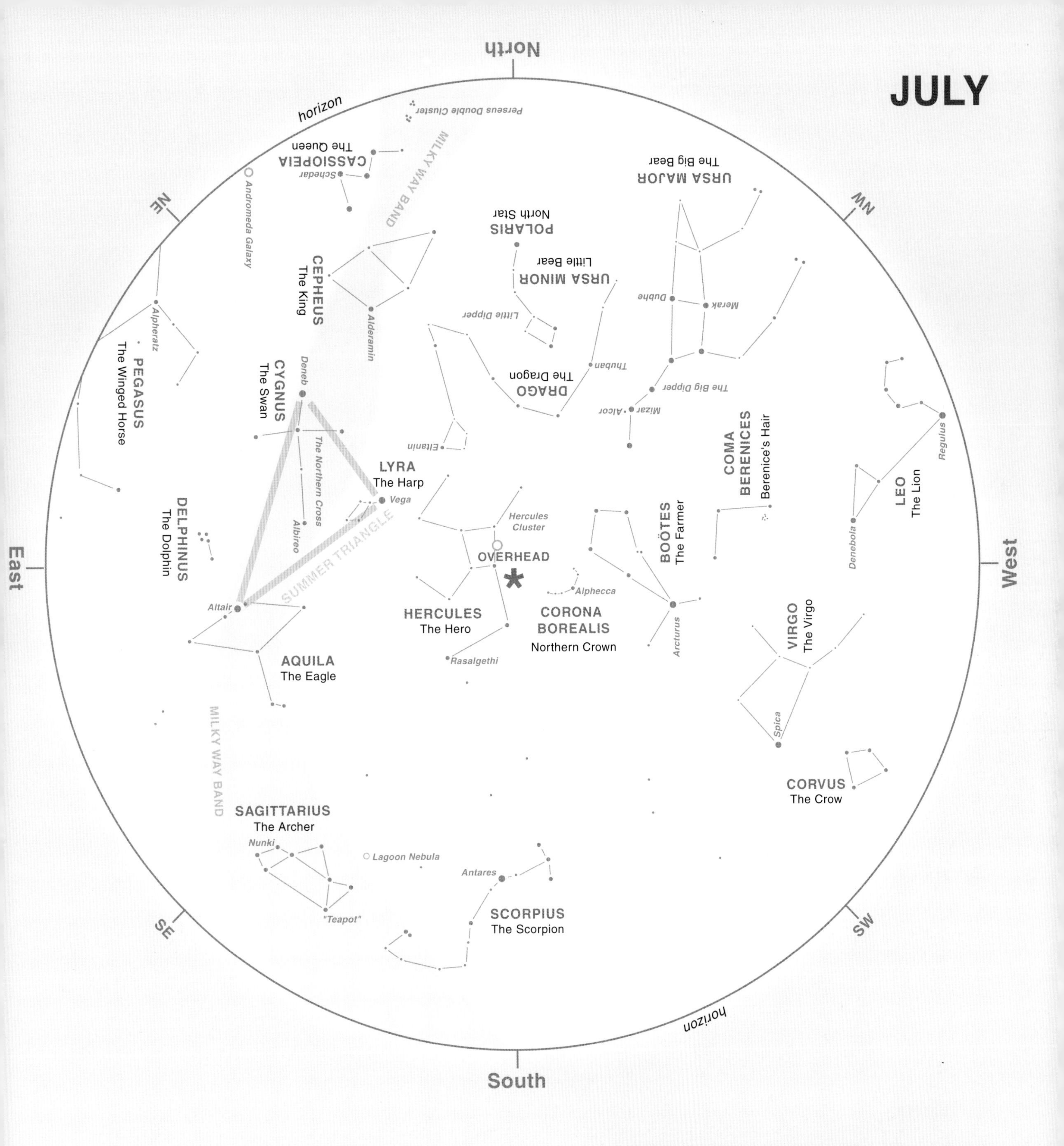

Use this map at the following times:

10 P.M. IN JULY
2 A.M. IN MAY
5 A.M. IN MARCH

AUGUST

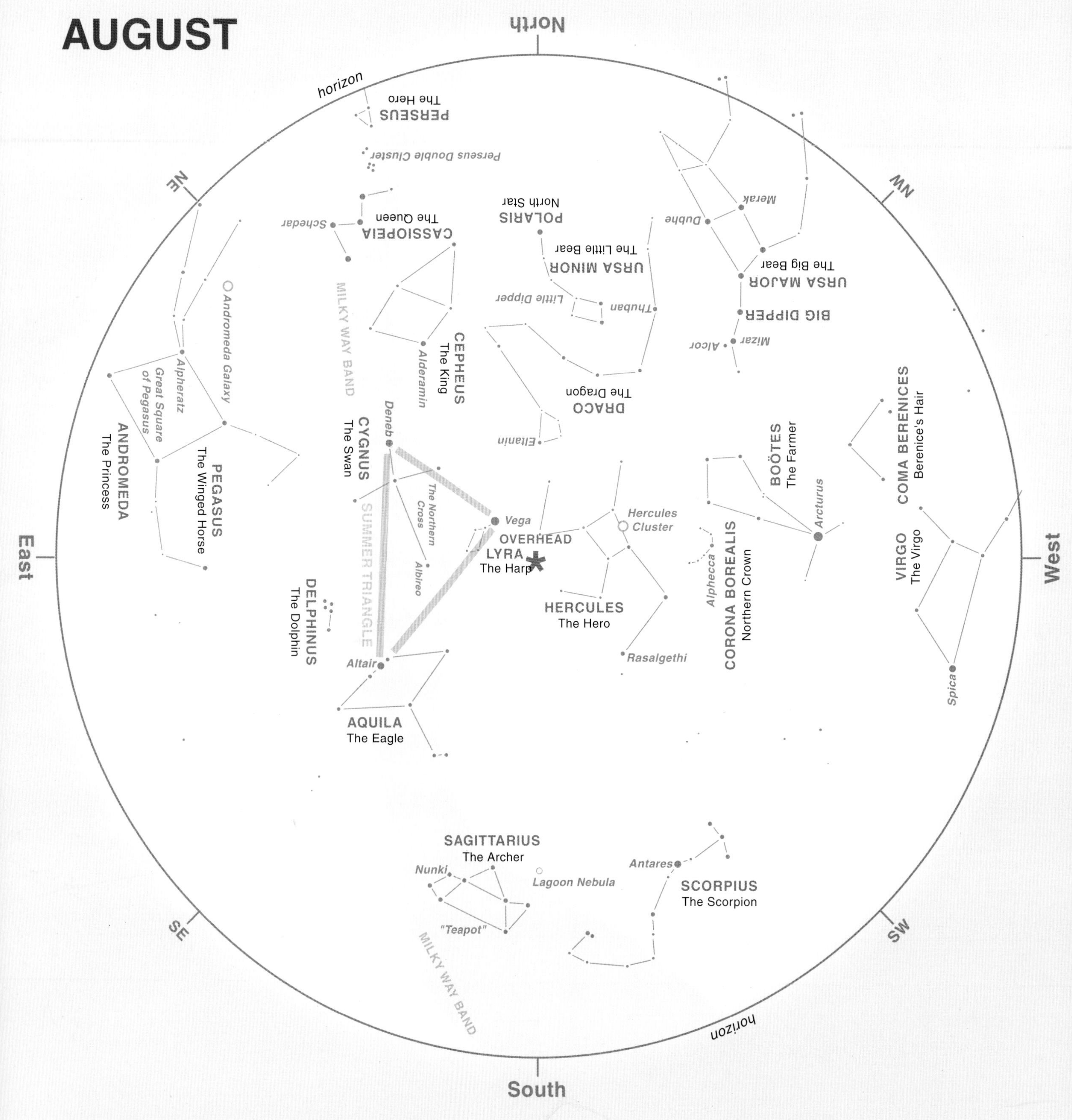

Use this map at the following times:

10 P.M. IN AUGUST
1 A.M. IN JUNE
4 A.M. IN APRIL

SEPTEMBER

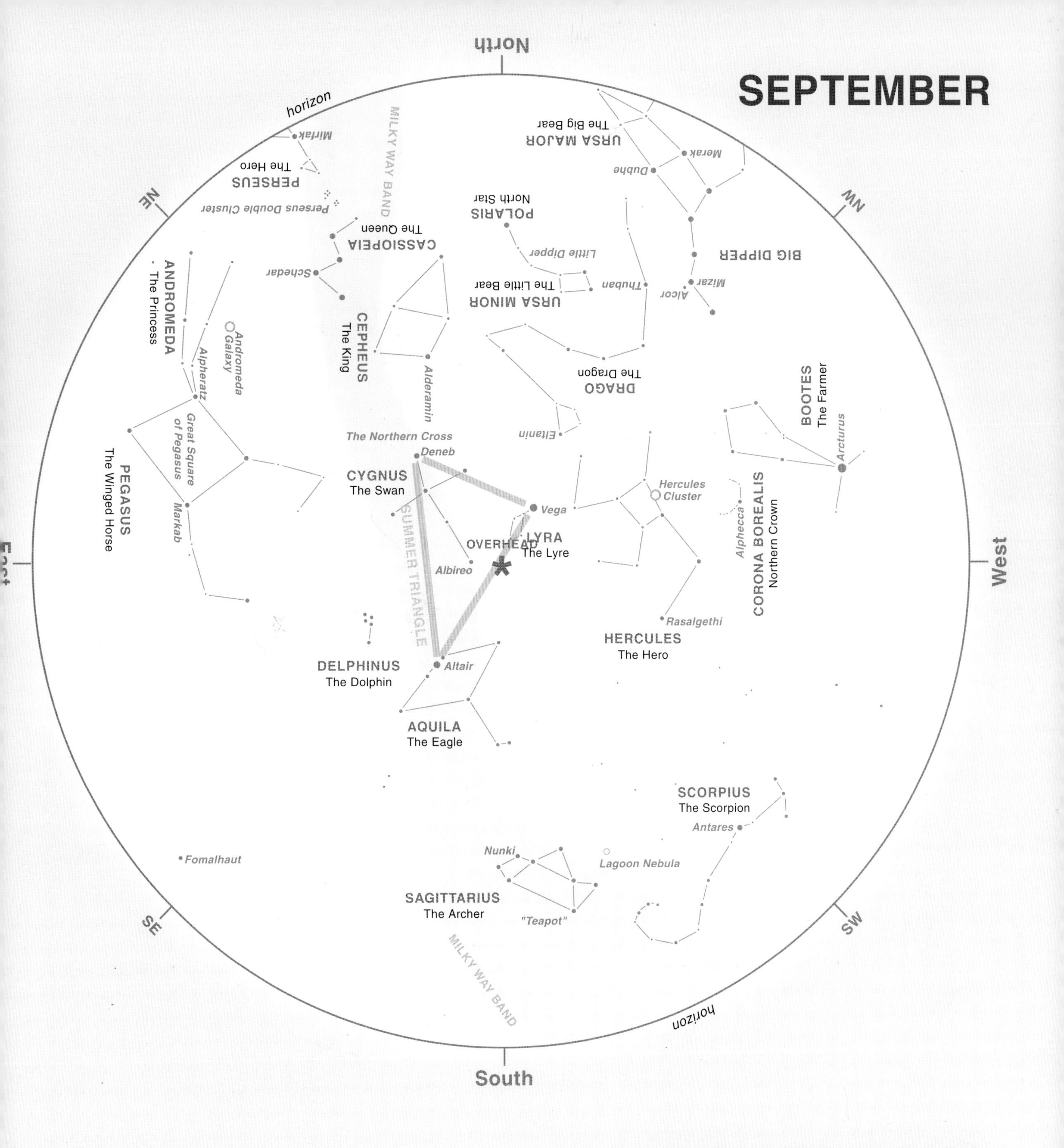

Use this map at the following times:

9 P.M. IN SEPTEMBER
1 A.M. IN JULY
4 A.M. IN MAY

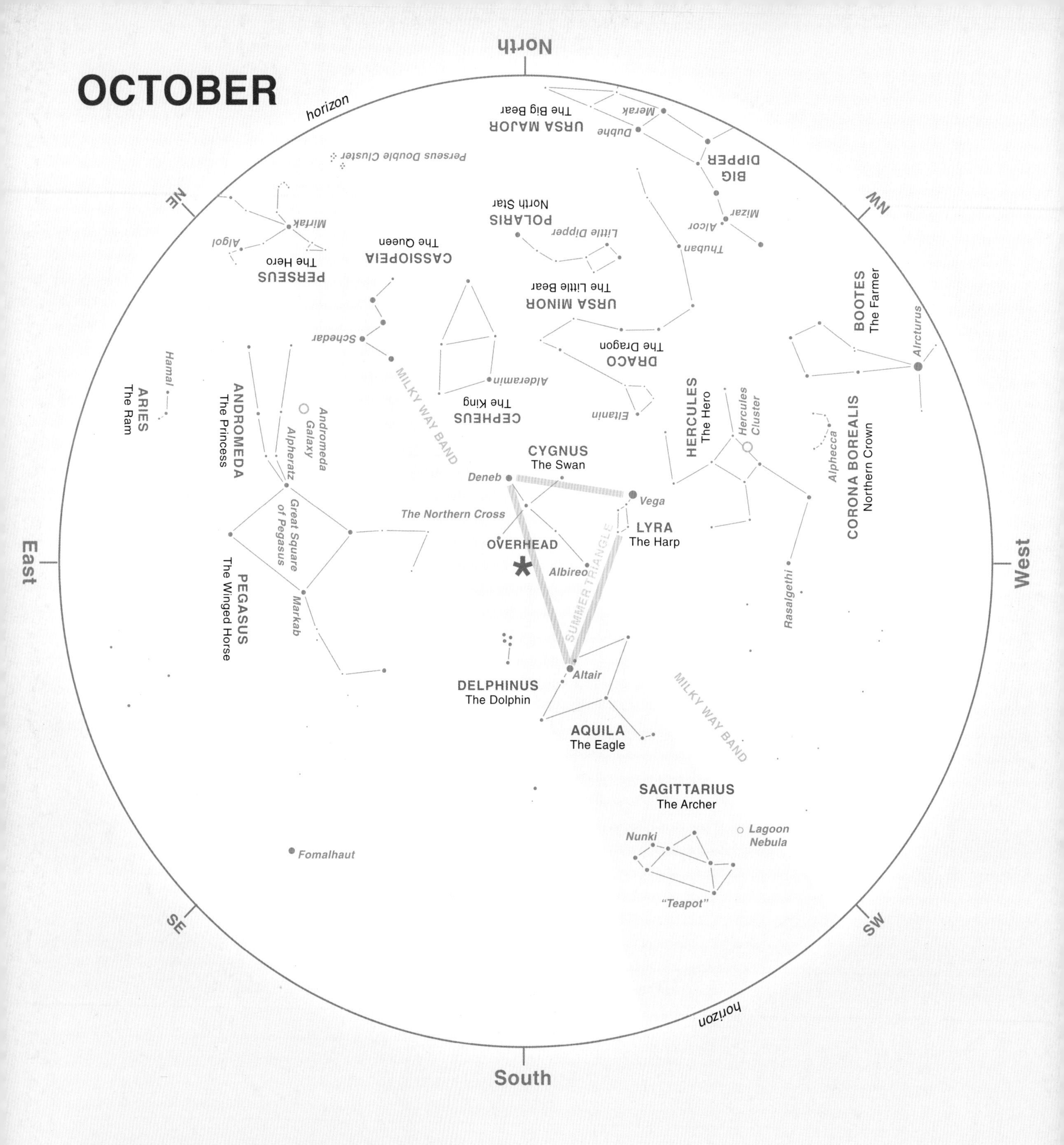

Use this map at the following times:

8 P.M. IN OCTOBER
MIDNIGHT IN AUGUST
3 A.M. IN JUNE

NOVEMBER

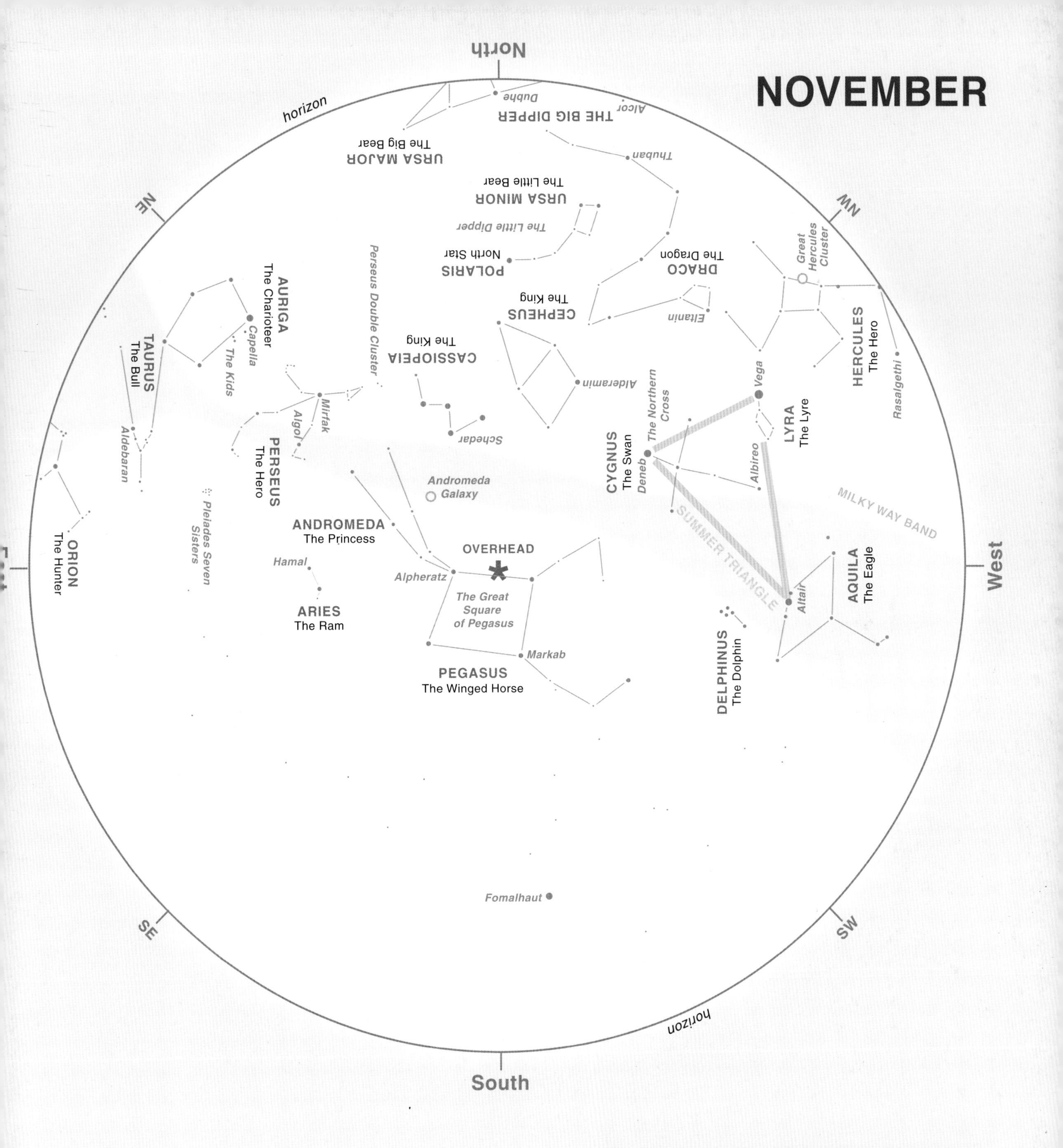

Use this map at the following times:

8 P.M. IN NOVEMBER
MIDNIGHT IN SEPTEMBER
3 A.M. IN JULY

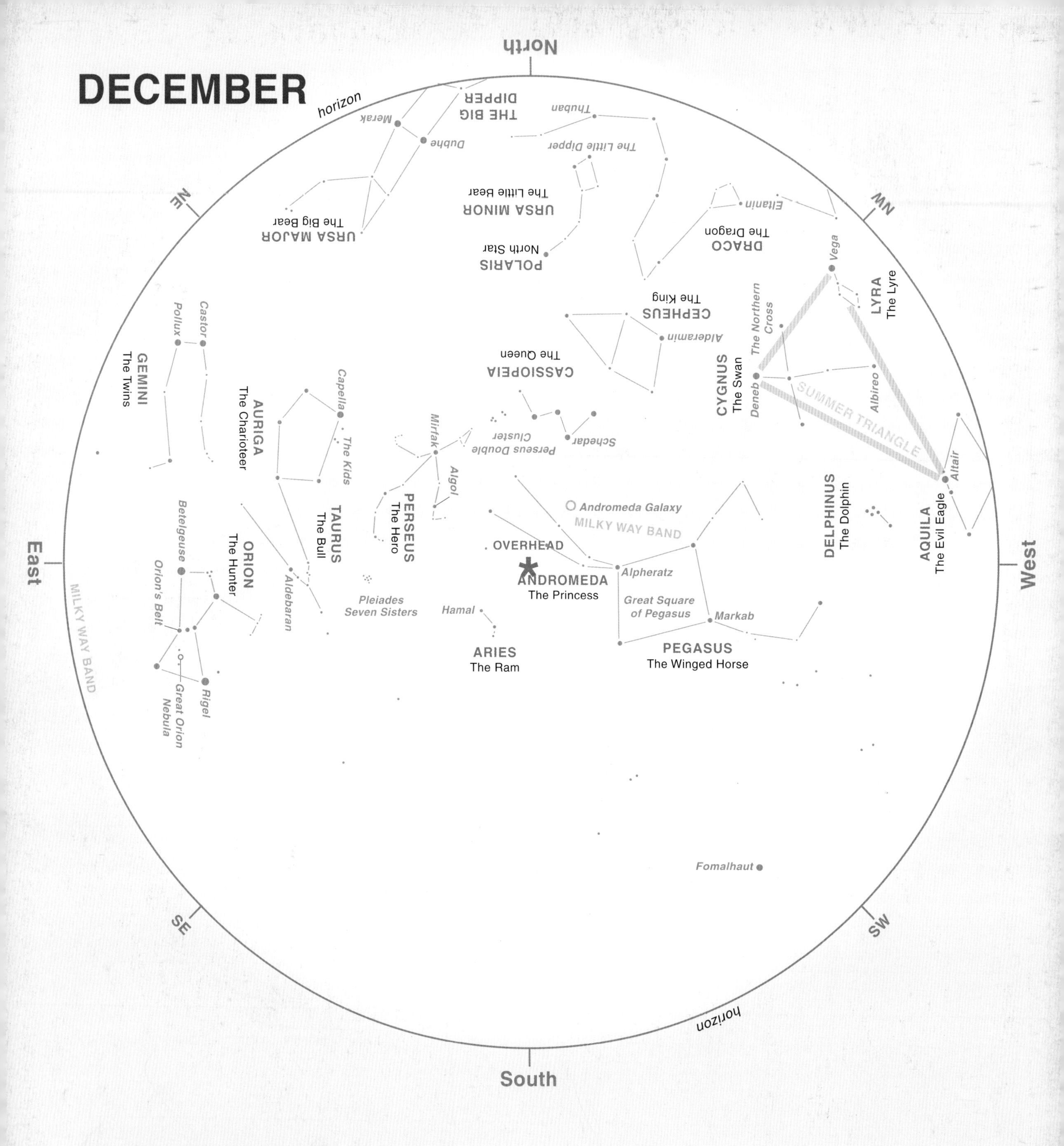

Use this map at the following times:

8 P.M. IN DECEMBER
MIDNIGHT IN OCTOBER
4 A.M. IN AUGUST